ISBN 978-1-5278-1636-7
PIBN 10909179

1 MONTH OF
FREE
READING

at
www.ForgottenBooks.com

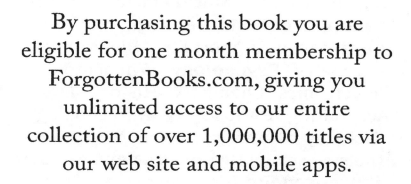

By purchasing this book you are eligible for one month membership to ForgottenBooks.com, giving you unlimited access to our entire collection of over 1,000,000 titles via our web site and mobile apps.

To claim your free month visit:
www.forgottenbooks.com/free909179

U.S. Army Coast. Eng. Res. Ctr. Tech. Rep. **CERC-83-1**

TECHNICAL REPORT CERC-83-1

SHORELINE MOVEMENTS

Report 1

CAPE HENRY, VIRGINIA, TO
CAPE HATTERAS, NORTH CAROLINA, 1849-1980

by

Craig H. Everts

Coastal Engineering Research Center
U. S. Army Engineer Waterways Experiment Station
P. O. Box 631, Vicksburg, Miss. 39180

and

Jeter P. Battley, Jr., and Peter N. Gibson

National Ocean Service
National Oceanic and Atmospheric Administration
U. S. Department of Commerce
6001 Executive Blvd., Rockville, Md. 20852

July 1983

Report 1 of a Series

Destroy this report when no longer needed. Do not return
it to the originator.

The findings in this report are not to be construed as an official
Department of the Army position unless so designated
by other authorized documents.

The contents of this report are not to be used for
advertising, publication, or promotional purposes.
Citation of trade names does not constitute an
official endorsement or approval of the use of
such commercial products.

REPORT DOCUMENTATION PAGE		READ INSTRUCTIONS BEFORE COMPLETING FORM
1. REPORT NUMBER Technical Report CERC-83-1	2. GOVT ACCESSION NO.	3. RECIPIENT'S CATALOG NUMBER
4. TITLE *(and Subtitle)* SHORELINE MOVEMENTS; Report 1: CAPE HENRY, VIRGINIA, TO CAPE HATTERAS, NORTH CAROLINA, 1849-1980		5. TYPE OF REPORT & PERIOD COVERED Report 1 of a series
		6. PERFORMING ORG. REPORT NUMBER
7. AUTHOR*(s)* Craig H. Everts Jeter P. Battley, Jr. Peter N. Gibson		8. CONTRACT OR GRANT NUMBER*(s)*
9. PERFORMING ORGANIZATION NAME AND ADDRESS U. S. Army Engineer Waterways Experiment Station Coastal Engineering Research Center P. O. Box 631, Vicksburg, Miss. 39180 and U. S. Department of Commerce National Oceanic and Atmospheric Administration National Ocean Service 6001 Executive Blvd., Rockville, Md. 20852		10. PROGRAM ELEMENT, PROJECT, TASK AREA & WORK UNIT NUMBERS
		12. REPORT DATE July 1983
		13. NUMBER OF PAGES 113
11. CONTROLLING OFFICE NAME AND ADDRESS Office, Chief of Engineers, U. S. Army Washington, D. C. 20314 National Oceanic and Atmospheric Administration 6001 Executive Blvd., Rockville, Md. 20852		15. SECURITY CLASS. *(of this report)* Unclassified
		15a. DECLASSIFICATION/DOWNGRADING SCHEDULE
14. MONITORING AGENCY NAME & ADDRESS*(if different from Controlling Office)*		

16. DISTRIBUTION STATEMENT *(of this Report)*

Approved for public release; distribution unlimited.

17. DISTRIBUTION STATEMENT *(of the abstract entered in Block 20, if different from Report)*

18. SUPPLEMENTARY NOTES

Available from National Technical Information Service, 5285 Port Royal Road, Springfield, Va. 22161.

19. KEY WORDS *(Continue on reverse side if necessary and identify by block number)*

Atlantic coast
Beach erosion
Coastal morphology
Shorelines

20. ABSTRACT *(Continue on reverse side if necessary and identify by block number)*

Shoreline position changes between about 1850 and 1980 along the ocean coastal reach from 12 km west of Cape Henry, Virginia, to 8 km west of Cape Hatteras, North Carolina, are documented in this report. In places where the ocean shoreline is on an island or spit, shoreline changes in the sound or bay are also given. Shoreline movement maps at a scale of 1:24,000 constitute the basic data set included in the report. Composite reproductions of these maps are shrinkwrapped separately. In addition, ocean and sound shoreline changes
(Continued)

DD FORM 1473 1 JAN 73   EDITION OF 1 NOV 65 IS OBSOLETE

20. ABSTRACT (Continued).

averaged for 1-minute-latitude- (or longitude-) distance increments are provided.

Consistent alongshore trends in the shoreline change rate are evident only from Virginia Beach south to the Virginia-North Carolina border and for about 15 km north of Cape Hatteras. Other areas experienced variable rates of shoreline change. The highest average shoreline change rate, about -2.0 m/year, occurred between 1917 and 1949. From about 1850 to 1917, the shoreline change rate averaged -0.1 m/year, and for the past 30 years it has averaged about -0.8 m/year.

PREFACE

This report is the result of a cooperative effort of the National Ocean Service (NOS), National Oceanic and Atmospheric Administration, U. S. Department of Commerce, and the Coastal Engineering Research Center (CERC) of the U. S. Army Engineer Waterways Experiment Station (WES). The study, based on a comparison of historic survey data contained in the archives of NOS, was funded jointly by the Office, Chief of Engineers, and the National Oceanic and Atmospheric Administration. All survey data reduction and quality control were performed by NOS; data analyses and report preparation were accomplished primarily by CERC.

The report was prepared by Dr. Craig H. Everts, CERC, and Messrs. Jeter P. Battley, Jr., and Peter N. Gibson, NOS. The work was carried out under the general supervision of Mr. N. E. Parker, Chief, Engineering Development Division, CERC; Mr. R. P. Savage, Chief, Research Division, CERC; and Dr. R. W. Whalin, Chief, CERC. At CERC, Mr. Edward Hands developed a computer program to analyze shoreline change data and Mr. Jon Berg reduced the data. The section on historic inlets was researched by Ms. Marie Ferland, CERC. Reviewers included Drs. Robert Byrne and Robert Dolan and Messrs. William Birkmeier, Edward Hands, Thomas Jarrett, James Melchor, Neill Parker, and S. Jeffress Williams.

Commander and Director of WES during the publication of this report was COL Tilford C. Creel, CE. Technical Director was Mr. F. R. Brown.

CONTENTS

Page

PREFACE   1

LIST OF TABLES   4

LIST OF FIGURES   4

CONVERSION FACTORS, INCH-POUND TO METRIC (SI)
  UNITS OF MEASUREMENT.   8

PART I: INTRODUCTION   9

PART II: STUDY AREA. 12

     Geographical Setting 12
     Historic Inlets 16
          Inlet location 16
          Accuracy of inlet location. 25
     Continental Shelf 26
     Tides, Winds, and Waves 29
          Tides and other sea level fluctuations. 29
          Wind conditions 32
          Waves 32
     Coastal Storms. 36
     Coastal Structures 38

PART III: DATA REDUCTION 44

     Data Sources. 44
     Shoreline Definition. 44
     Methods Used to Revise the 1980 Mean High Water Line 50
     Data Reduction Procedures 51
     Quality Control and Potential Errors 52

PART IV: DATA ANALYSIS AND DISCUSSION 55

     Analysis Methodology. 55
     Shoreline Change Rates 58
          Listing of shoreline change rates 58
          Ocean shoreline change rates. 62
          Sound shoreline change rates. 71
          Oregon Inlet. 71
          Cape Hatteras and Cape Henry. 78
          Variation in shoreline change rates with time 78
          Changes in island width and position 84

PART V: PREDICTION OF FUTURE SHORELINE CHANGES 92

     Temporal Predictions 92
     Spatial Predictions 92
          Barrier island migration and narrowing 93
          Alongshore sediment transport reversal. 96
          Sound shoreline change. 97
          Inlets and shore erosion 99
          Capes and shoreline change. 102
          Shoreface-connected ridges and shoreline change 103

                                                                      Page
PART VI: SUMMARY AND CONCLUSIONS   106
REFERENCES   109

NOAA/NOS shoreline movement maps, 1852-1980, are shrinkwrapped as a separate enclosure to this report.

LIST OF TABLES

No.   Page

1   References to Maps and Charts Used to Establish Historic
     Inlet Locations   18

2   History of Rudee Inlet, Virginia   20

3   Coastal Structures, Cape Henry to Cape Hatteras.   41

4   Historic Shoreline Surveys, Cape Henry to
     Cape Hatteras, 1847-1980.   46

5   Ocean Shoreline Changes in Virginia West of Cape Henry   59

6   Ocean Shoreline Changes in Virginia South of Cape Henry.   59

7   Ocean Shoreline Changes in North Carolina North of
     Cape Hatteras   60

8   Ocean Shoreline Changes West of Cape Hatteras.   61

9   Sound Shoreline Changes, Cape Henry to Cape Hatteras   63

10   Pamlico Sound Shoreline Changes West of Cape Hatteras.   64

11   Summary of Mean Shoreline Changes, Oceanside   82

12   Summary of Mean Shoreline Changes, Soundside   82

13   Combined Ocean- and Soundside Shoreline Changes.   85

LIST OF FIGURES

1   Location map   10

2   Cape Hatteras viewed toward the northwest   13

3   Frontal dune along the Atlantic Ocean side of Hatteras Island. . . 13

4   Jockey's Ridge at Nags Head, N. C., rises almost 50 m 14

5   Sand dunes encroaching on cottages and a forest at Kill
     Devil Hills, N. C.   14

6   Pea Island, N. C., viewed south across Oregon Inlet 15

7   Relic beach ridges at Cape Henry, Va.   15

8   View north across Oregon Inlet   16

9   Locations of persistent inlets reported open on maps and charts
     between Cape Hatteras and Cape Henry from 1585 to 1980   17

10   Inlet channel, possible related to the now-closed Trinity Harbor
     Inlet, in Currituck Sound west of the CERC field research
     facility   23

11   Barrier Island width versus duration of time inlets were open
     after 1585 between Cape Hatteras and Cape Henry   24

12   Probable site of a large pre-1585 inlet at Kitty Hawk, N. C. . . . 27

No.   Page

13   Continental shell profiles taken between Virginia Beach, Va.,
        and Hatteras Island, N. C., to 30 km from shore   28

14   Tide ranges, Cape Henry to Cape Hatteras   30

15   Tide frequencies for ocean shoreline at Kitty Hawk, N. C., for
        several classes of storms: (a) landfalling, (b) alongshore,
        (c) inland, (d) exiting hurricanes, (e) winter storms,
        (f) all storms   31

16   Surface wind roses, Cape Henry and vicinity, from data
        collected 1850-1960   33

17   Surface wind roses, Cape Hatteras and vicinity, from
        data collected 1850-1960   34

18   Annual cumulative significant wave height distribution based
        on 20 years of hindcast data measured at 10-m water depth
        off Kitty Hawk, N. C.   35

19   Wave rose diagram showing the significant wave height and direc-
        tion of wave propagation for combined 20-year hindcast data in
        10-m water depth at station 81 off Kitty Hawk, N. C.   36

20   Yearly storm surge return period for extratropical storms at
        Hampton Roads, Va.   37

21   Number of tropical cyclones reaching the North Carolina
        coast, by sector, for the period 1886-1970   38

22   Recreational Pier at Virginia Beach, Va., a typical fishing
        pier for the study area   39

23   View toward north of groins near Cape Hatteras Lighthouse   40

24   Weir jetty system at Rudee Inlet, Virginia Beach, Va.   42

25   U. S. Geological Survey 1:24,000 quadrangles used as base
        maps in the study   45

26   Digitization procedure for correcting shoreline position loca-
        tion when original shoreline movement map distortions exist . . .   52

27   Definition sketch illustrating parameters used to obtain
        shoreline change rates for a north-south-trending ocean
        shoreline   56

28   Ocean shoreline change rates from near Cape Henry to Cape
        Hatteras, from about 1850 to 1980   65

29   Standard deviation of ocean shoreline position changes between
        Cape Henry and Cape Hatteras   66

30   Average ocean shoreline change rates for the 36-km-long reach
        south of Cape Henry in the periods 1859-1925 and 1925-1980 . . .   67

31   Average ocean shoreline change rates for two survey periods in
        the reach between Duck, N. C., and Cape Hatteras   68

No.   Page

32   Average ocean shoreline change rates at Virginia Beach for four
      successive surveys between 1925 and 1980: (a) 1925-1942(44),
      (b) 1942(44)-1962, (c) 1962-1980   69

33   Extreme ocean shoreline excursions from about 1850 to 1980,
      Cape Henry to Cape Hatteras   70

34   Sound shoreline change rates for the reach between Back Bay, Va.,
      and 12 km west of Cape Hatteras   72

35   Standard deviation of sound shoreline position changes between
      Cape Henry and Cape Hatteras   73

36   Average sound shoreline change rates, for the periods from about
      1850 to 1915 and from about 1915 to 1980, between Nags Head
      and Cape Hatteras, N. C.   74

37   Changes in ocean and sound shorelines adjacent to Oregon Inlet,
      N. C., for five surveys between 1852 and 1980   75

38   Migration rates of Oregon Inlet throat for five survey intervals
      between 1849 and 1980   76

39   Relative locations and orientations of the narrowest section
      of the Oregon Inlet throat, 1849-1980   77

40   Plan view changes in land area in the vicinity of Oregon Inlet,
      N. C., 1849-1980   77

41   Changes in mean high-water shoreline at Cape Point, Cape
      Hatteras, between 1852 and 1980   79

42   Relative plan area of the subaerial projection for Cape
      Hatteras, 1852-1980   80

43   Location of Cape Hatteras point between 1852 and 1980   80

44   Changes in mean high-water shoreline at Cape Henry between
      1852 and 1980   81

45   Shoreline change rates, averaged by survey period, for east-
      facing ocean shorelines and west-facing sound shorelines 83

46   Shoreline change rates, averaged by survey period, for west-
      facing ocean and sound shorelines   83

47   Island width changes between about 1850 and 1980, from Cape
      Henry to west of Cape Hatteras   87

48   Island width change rates, 1852-1917 and 1917-1980, between
      Kitty Hawk and Cape Hatteras, N. C.   88

49   Rates of position change of island axis between about 1850 to
      1980, Cape Henry to west of Cape Hatteras   89

50   Rates of position change of island axis, 1852-1917 and
      1917-1980, between Kitty Hawk and Cape Hatteras, N. C.   90

51   Relationship of apparent ocean shoreline change to Capes,
      shoreface-connected ridges, and Oregon Inlet   94

No.   Page

52   Ocean shoreline changes from about 1850 to 1980, Cape Henry
      to Cape Hatteras, as a function of island width in 1980   98

53   Sound shoreline changes from about 1850 to 1980, Cape Henry to
      Cape Hatteras, as a function of island width in 1980   99

54   Bathymetry seaward of the ocean shore between Cape Henry and
      Cape Hatteras 104

55   Shoreline orientation referenced to true north between Cape
      Henry and Cape Hatteras 105

CONVERSION FACTORS, INCH-POUND TO METRIC (SI)
UNITS OF MEASUREMENT

Inch-pound units of measurement used in this report can be converted to metric (SI) units as follows:

Multiply	By	To Obtain
cubic yards	0.7645549	cubic meters
feet	0.3048	meters
inches	0.0254	meters
knots (international)	0.514444	meters per second
miles (U. S. statute)	1.609347	kilometers
miles per hour	1.609347	kilometers per hour

SHORELINE MOVEMENTS

Report 1

CAPE HENRY, VIRGINIA, TO CAPE HATTERAS, NORTH CAROLINA, 1849-1980

PART I: INTRODUCTION

1. This report describes results of a cooperative National Oceanic and
Atmospheric Administration (NOAA), National Ocean Service (NOS), and U. S.
Army Engineer Waterways Experiment Station, Coastal Engineering Research
Center (CERC), study of shoreline changes. The study area comprises the ocean
coast south from Cape Henry, Virginia, to west of Cape Hatteras, North Caro-
lina, and the sound-side coast of the barrier islands between each of the
Capes (Figure 1). Changes in shoreline position from 1852 to 1980 are treated
using survey data from NOS and its predecessor, the U. S. Coast and Geodetic
Survey (C&GS). (NOAA/NOS-CERC shoreline movement maps, 1852-1980, are included as a separate enclosure to this report.)

2. Shoreline changes of a quantifiable nature are presented covering
what is probably the longest period of historic survey record of the area
available. Although maps exist dating back to 1585 (Cumming 1966), prior to
1849 the position of the shoreline was not located with sufficient accuracy to
allow a comparison of that feature on different maps. The early maps, however,
do provide a valuable reference for locating inlets that were open during the
past 400 years. Langfelder et al. (1970), in a study of coastal erosion in
North Carolina, used aerial photographs dating from 1945, for which measurements were made at approximately 300-m intervals along the beach. Dolan
et al. (1979) also using aerial photographs but measuring at 100-m intervals,
established erosion rates in Virginia, North Carolina, and elsewhere, based
upon data spanning 30 years or more for over half the area and over 15 years
for the whole area. Dolan et al. (1979, p 603) note their total measurement
error as potentially as much as ±25 m for rate-of-change calculations. The
frequency of the aerial survey was much greater than that of shoreline surveys
used in this study, but the total aerial study duration was less than 25 percent that of this study. This longer data span (130 years) allows a more
extended analysis of temporal variations in shoreline change rates.

9

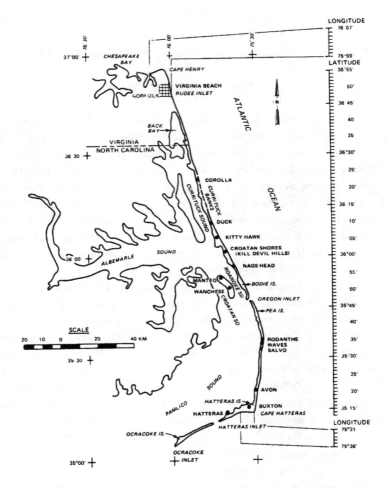

Figure 1. Location map (the study area shoreline is shown as
a heavy line; data are not available where the heavy line is
dashed. The northwest boundary of the study area is 12 km
west of Cape Henry at the Chesapeake Bay entrance; the southwest boundary is 8 km west of Cape Hatteras)

10

3. This report provides a long-term basic data set for use in management and engineering decisions related to the coastal zone. In the absence of other data, past shoreline changes usually provide the best available basis for predicting future changes. An extrapolation of past changes is not without risk, though. Man's actions may have affected the natural coastal change processes and thereby altered the rates of change. Probably more importantly, the material processes themselves may have altered over time thereby varying the shoreline change rate; Hayden (1975), for example, has identified relatively large changes in storm-wave climate in this century at Cape Hatteras.

4. Historic shoreline change data are direct, believable, and explicit and can be updated as new data become available. Shoreline changes obtained from historic charts for a specific time period also are invariant. Past shoreline changes based on NOS surveys can be supported in a court of law.

5. Coastal engineers use past shoreline changes in the design of projects for shoreline stabilization, flood prevention as a result of storm surges, and maintenance of navigable depths in coastal waterways. A knowledge of past changes in shoreline position is a useful and often necessary basis from which to predict the effects of natural processes and proposed modifications on the coastal zone.

6. This is an empirical report. It serves to explain and enhance the shoreline change maps which go with it. Since it is sometimes difficult to determine trends from maps alone, average changes have been calculated for each minute of latitude (north-south-trending shoreline) and longitude (east-west-trending shoreline). Relationships are established between the shoreline change rates and (a) shore orientation, (b) location of capes, (c) proximity to present inlets and inlets that were historically open, (d) shore-connected ridges, and (e) an alongshore sediment transport nodal reach. A brief description of wind, wave, tide, and sedimentological parameters in the study area is provided in Part II for readers interested in those factors; however, because their records are insufficiently detailed or too short with respect to shoreline changes, these parameters are not used further in this report.

PART II: · STUDY AREA

Geographical Setting

7. The study area encompasses 210 km of Atlantic Ocean barrier island coast. It begins in the north 12 km west of Cape Henry, Virginia, and extends south to 8 km west of Cape Hatteras, North Carolina (Figure 1). A bay and four sounds back the barrier islands along the southern 175 km. of ocean shore. These include Back Bay, Currituck Sound, Albermarle Sound, Roanoke Sound, and Pamlico Sound. Presently, only Oregon Inlet connects a sound and the ocean in the study area. Rudee Inlet provides ocean access from a small lake near Virginia Beach.

8. Currituck Banks now extends south from Back Bay, Virginia, to Oregon Inlet, North Carolina. A past segment of the Banks from the vicinity of Kitty Hawk, North Carolina, to Oregon Inlet is still called Bodie Island. Beyond Oregon Inlet, the barrier is known as Pea Island about as far south as Rodanthe, North Carolina, and as Hatteras Island from Rodanthe to Hatteras Inlet, North Carolina; the boundary between the two lies at the site of now-closed New Inlet. Hatteras Island is sharply angled to the southwest at Cape Hatteras. The Cape is one of the most conspicuous cuspate headlands along the Atlantic Coast (Figure 2).

9. The barrier islands vary in width from 0.5 to almost 5 km. A frontal dune backs most of the barrier beach (Figure 3). Dunes west of the frontal dune, most notably Jockey's Ridge, North Carolina (Figure 4), also are found along some sections. Hennigar (1979) found these dunes to be moving to the southwest at Kill Devil Hills, North Carolina (Figure 5), and elsewhere. In most locations, aeolian, overwash, and relict flood-tidal delta flats extend from the dunes to the sound (Figure 6). Relic beach ridges exist in the flats area at Kitty Hawk, west of Cape Hatteras, and at Cape Henry (Figure 7).

10. Sand size varies in an alongshore direction, across the beach and from season to season. From the Virginia-North Carolina line to Cape Hatteras, the median foreshore sand size is 0.44 mm, with a slight average increase from north to south (Shideler 1973). Within this area, the beach from between Corolla and Duck to Kitty Hawk, North Carolina, is composed of anomalously large, iron-stained quartz and feldspar sand in the 1-mm-diameter range. Beach sand north of the States boundary is finer. Average dune sand size in

12

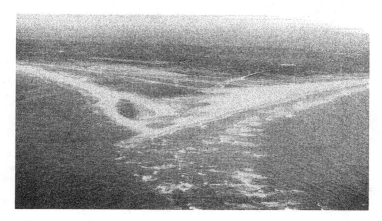

Figure 2. Cape Hatteras viewed toward the northwest (the Atlantic Ocean is in the foreground; Pamlico Sound is in the background)

Figure 3. Frontal dune along the Atlantic Ocean side of Hatteras Island between Salvo and Avon, N. C.

Figure 4. Jockey's Ridge at Nags Head, N. C., rises almost 50 m (the Atlantic Ocean shore is in the foreground; Albemarle Sound is in the background) (Hennigar 1979)

Figure 5. Sand dune encroaching on cottages and a forest at Kill Devil Hills, N. C. (dune movement is to the southwest; i.e., toward the left background of the photograph)

Figure 6. Pea Island, N. C., viewed south across Oregon Inlet (overwash and flood-tide delta flats comprise most of the western two-thirds of the island; the Atlantic Ocean is at the left; Pamlico Sound is at the right of this photograph)

Figure 7. Relic beach ridges at Cape Henry, Va. (these ridges formed in the past as the cape built north and eastward; Virginia Beach is at the foreground)

15

the study area is 0.27 mm and does not vary from north to south.

Historic Inlets

11. Inlets have played and continue to play an important role in shoreline evolution in the study area. At present only two inlets, Rudee and
Oregon, are open; in the past as many as seven have been open simultaneously.
The inlets act as traps for littoral sediments which move into the lagoons
from adjacent ocean beaches and in this way contribute to ocean shoreline retreat. The sound shoreline is often moved toward the mainland by sand accretion in flood-tidal deposits behind the islands and adjacent to open inlets
(Figure 8). Closed inlet locations are frequently distinguishable by a bulge
in the sound shoreline.

Inlet location

12. Figure 9 shows the extent of inlets reported open since 1585 on

Figure 8. View north across Oregon Inlet (most of the
large shoreline lobes and islands in Pamlico Sound at
the left are relic flood-tidal shoals and other inlet
features created as Oregon Inlet migrated south; the
sound shoreline, therefore, moved west as the inlet
trapped beach sand)

16

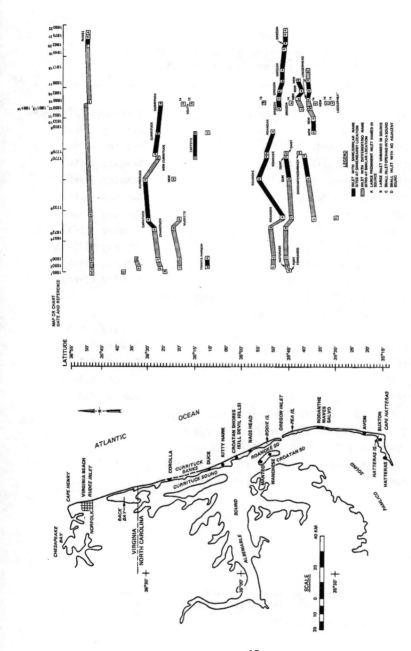

Figure 9. Location of persistent inlets reported open on maps and charts between Cape Hatteras and Cape Henry from 1585 to 1980 (Table 1 lists map and chart dates and secondary references; inlet names are shown here as they appear in the sources. Inlets with solitary boxed letters (see Legend) appear to have been isolated in time and location; solid and stippled bars (see Legend) show time connections between open inlets in consecutive sources) (Caffeys Inlet data from Fisher (1962))

17

Table 1

References to Maps and Charts Used to Establish
Historic Inlet Location (Figure 9)

Reference Number (Fig. 9)	Date	Author	Secondary Reference
1	1585	White	Cumming (1966)
2	1590	White-DeBry	Cumming (1966)
3	1606	Mercator-Flordius	Cumming (1966)
4	1657	Comberford	Cumming (1966)
5	1672	Ogilby-Moxon	Cumming (1966)
6	1733	Moseley	Cumming (1966)
7	1770	Collet	Cumming (1966)
8	1775	Mouzon	Cumming (1966)
9	1808	Price-Strother	Cumming (1966)
10	1833	MacRae-Brazier	Cumming (1966)
11	1852	NOAA/NOS-CERC shoreline change maps*	
12	1859	NOAA/NOS-CERC shoreline change maps	
13**	1861	Bachman	Cumming (1966)
14**	1861	Colton	Cumming (1966)
15	1865	U. S. Coast Survey	Cumming (1966)
16	1882	Kerr-Cain	Cumming (1966)
17**	1896	Post Route Map	Cumming (1966)
18	1917	NOAA/NOS-CERC shoreline change maps	
19	1949	NOAA/NOS-CERC shoreline change maps	
20	1962	NOAA/NOS-CERC shoreline change maps	
21	1975	NOAA/NOS-CERC shoreline change maps	
22	1980	NOAA/NOS-CERC shoreline change maps	

*   Published as a separate inclusion to this report.
** Maps not discussed in Fisher (1962).

the maps and charts of the times found in Cumming (1966), on NOAA/NOS-CERC shoreline change maps dating from 1852, or, in the case of Caffeys Inlet, according to data collected by Fisher (1962). The following inlets warrant specific comment on their locations as shown in Figure 9.

13. Rudee Inlet. The inlet shown at approximately 36°48' in several of the very early maps (1585, 1590, and 1606) was located by position in relation to geomorphic features rather than by latitude, since latitude was less accurate for location purposes prior to the late 1700's. The inlet was possibly open in 1682 (Cumming 1966, figure on p 14); however, on a copy of a 1682 map "Rudee" was written next to a lake which has the same general configuration of Rudee Lake today. A history of Rudee Inlet after 1927 is given in Table 2.

14. "Back Bay" Inlet at latitude 36°33'-34' (1590, 1606). On the original maps, this inlet did not open into a large sound or bay but instead appeared as a small indentation in the coastline. Comparing geomorphological features and the mainland shoreline shows that this inlet sequence actually existed just south of Back Bay, opposite Knotts Island. It was most likely the precursor to Old Currituck Inlet which was shown in later years as having closed at approximately this location (1833, 1861 (Colton), 1865, 1882); the 1882 map states that Old Currituck Inlet closed in 1775.

15. Caffeys Inlet at latitude 36°15'. Early mention of this inlet in a report by the North Carolina Fisheries Commission Board (1923, p 33) shows that the inlet was open for a short time between 1780 and 1800. The location can be deduced from the text to be south of Currituck Inlet, but no map was included in the report.

16. Dunbar (1958, p 218) placed the inlet at approximately 36°13', calling it Carthys Inlet and showing it open from at least 1798 to 1811. He concluded that the inlet opened at the site of Trinity(e) Harbor (1585-?) and that the same location was later called South Inlet (1808, 1833, 1861), though the inlet had actually closed by that time.

17. Fisher (1962, p 90) shows Caffeys Inlet to be north of the town of Duck at 36°15' and open from 1770 to 1811, maximum. He bases this location on the existence of a large, relict, flood-tidal delta feature which he felt was a more likely site than the relatively narrow segment of the island at 36°13' where the Caffeys Inlet Coast Guard Station is now located. Fisher's location is shown in Figure 9.

18. The Price-Strother map of 1808 shows an unnamed inlet at 36°11',

Table 2

History of Rudee Inlet, Virginia (from U. S. Army
Engineer District, Norfolk (1982))

Date	Event
pre-1927	Shallow drainage ditch that opened and closed frequently
1927	Virginia Highway Department constructed a concrete culvert and built a highway over it
1933	Hurricane destroyed both the culvert and the highway
1933-1952	Inlet open but less than 18 in. deep (and meandering to some degree)
1952	Virginia Beach Erosion Commission organized
1953	Virginia Beach Erosion Commission constructed two short jetties on either side of the inlet and a sheet pile wall on north side
1954-1962	A fixed dredge was installed on the end of the south jetty to bypass sand
1962	"Ash Wednesday" storm destroyed the bypassing plant
1962-present	Small dredges have operated periodically with limited success. Several commercial dredging operations have also been completed to ±6 ft* mean low water (mlw) project depth
1968	Existing jetties were extended north, by 560 ft, and south, by 280 ft, in addition to a 477-ft-long timber weir. Also, a 100,000-cu yd sand trap was dredged to -16 ft
1975	Waterways Experiment Station (WES) installed a test jet-pump bypassing system
1975	Virginia Beach purchased the system from WES. This system was operating through 1982
1979-spring	A commercial dredge opened the filled sand trap and removed approximately 100,000 cu yd of material
1980-spring	A commercial dredge opened the sand trap and removed approximately 100,000 cu yd of material

* A table for converting the inch-pound units of measure in this report to metric (SI) units is found on page 8.

but placement by geomorphic features corresponding to current maps indicates that the latitude is actually 36°14'-15'. This is most likely the inlet known as Carthys and Caffeys (and perhaps South in 1861 maps).

19. <u>South Inlet at latitude 36°16'-18'</u>. Dunbar (1958, p 138) makes two references to South Inlet (1830, 1833); he states that the inlet had actually closed by the referenced time and was "...probably an example of cartographic perpetuation of a feature no longer in existence." He gives no reason for the change in name from Caffeys to South Inlet, though he considers them to be at the same location.

20. South Inlet appears at approximately 36°16'-17' on the 1861 maps that Cumming (1966) considered during his study. It is probably not significant that South Inlet appears on the Bachman map because the map is inaccurate. Colton also shows South Inlet on his 1861 map, but it is quite possible that all inlets on his map should be shifted to the north by approximately 5' of latitude; if South Inlet were shifted northward, it could be considered part of the Currituck Inlet system found between 36°26'-27' at that time. South Inlet is shown in Figure 9, but it may not represent a single event at that location.

21. <u>Trinity Harbor Inlet at latitude 36°12'</u>. Dunbar (1958, p 216) placed Trinity Harbor (1585-?) at approximately 36°13' and regarded it as the precursor to Carthys Inlet, now the site of Caffeys Inlet Coast Guard Station, which was open from at least 1798 to 1811. Interestingly, Dunbar's location of Caffeys Inlet is 1'-2' south of the large flood-tidal delta sequence at a narrow section of the barrier beach mentioned in paragraph 17.

22. Fisher (1962, p 110) discussed the location of Trinity Harbor and concluded that Dunbar's assumption of its location was incorrect because it would be unusual for an inlet to open on the site of an earlier inlet. He goes on to say that Trinity Harbor was most likely located further to the north at 36°17' where there is a relict inlet feature (presently called Beasley Bay).

23. The White-DeBry map of 1590 (Figure 9) shows Trinity Harbor to be north of the wide Kitty Hawk/Southern Shores feature, directly east of a small embayment and just south of an unnamed inlet with associated islands. Close examination of this 1590 map and comparison with current maps suggests that a location of 36°11'-12' is more accurate; Fisher's placement to the north by almost 5' of latitude seems to be based almost entirely on the relict inlet

21

feature. In addition to the 1590 and 1606 maps, strong evidence for the existence of Trinity Harbor Inlet at the more southerly location includes:

a. A channel of 5- to 7-ft depths (where adjacent water depths are 2-3 ft on the average) in Currituck Sound (Figure 10).

b. Inlet/channel fill sediments recorded (Field 1973) from cores taken when the CERC Field Research Facility was constructed.

c. A slight westward bulge in the Currituck Sound shoreline which could be the remnant of a reworked flood-tidal delta.

24. 1657 Comberford map. This map depicts two large unnamed inlets open in the stretch of coast between Currituck Sound and present-day Oregon Inlet. The two inlets extend for the equivalent of at least 5' and 2', respectively, of latitude fronting Roanoke Island for most of its length; this is probably a distortion, since earlier and later maps showed much narrower inlets. Placement of both of these class B inlets in Figure 9 (possibly Roanoke and Gunt) at the midpoint of the location listed on the 1657 map is subjective.

25. Kitty Hawk Bay region at latitude 36°00' to 36°15'. Three distinctive features suggest a prehistoric inlet in this region:

a. A wide "field" of long beach ridges (Figure 11), recurving and ending abruptly to the south at Kitty Hawk Bay, which could have been formed during the migration of an inlet.

b. Kitty Hawk Bay itself and the narrow section of the barrier island which separates the bay from the Atlantic Ocean.

c. Collington Island, a large feature composed of both sandy areas and salt marsh, which closely resembles a relict flood-tidal delta.

Early maps (1585, 1590, and 1606, to name the earliest) delineate this multiple feature quite clearly. Therefore, depositional processes that formed the feature were active before 1585, and the area has (approximately) maintained its present configuration through historic time.

26. Chacandepeco Inlet at latitude 35°16'-17'. In 1923, the North Carolina Fisheries Commission Board (1923, p 17) suggested that an inlet be opened 3 miles north of the Hatteras Lighthouse to increase the fishing potential of Pamlico Sound. This was considered to be an optimum location for an inlet because of (a) the existence of "Cape Channel," a deep channel in Pamlico Sound; (b) the narrowness of the island; and (c) the distance from another major inlet. Previous existence of an inlet, however, was not mentioned.

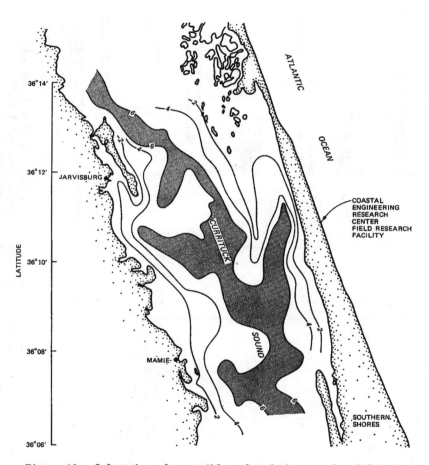

Figure 10. Inlet channel, possibly related the now-closed Trinity Harbor Inlet, in Currituck Sound west of the CERC Field Research Facility (contour lines show depth in feet)

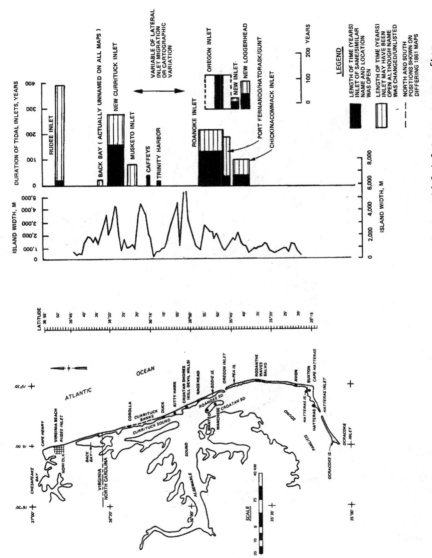

Figure 11. Barrier island width versus duration of time tidal inlets were open after 1585 between Cape Hatteras and Cape Henry (width was measured on NOAA 1:80,000 (scale) Nautical Charts 1227, 12204, and 11555

24

27. Dunbar (1958, p 217) stated that the inlet was open from 1585 to 1687 and called Chacandepeco by the Indians. He also noted that, although the inlet was shown on Comberford's 1657 map, it could have been copied from earlier maps and not actually open at that time.

28. Fisher (1962, p 92-93) concluded that an inlet was open from pre-1585 to 1672 at 36°16.5'; his conclusion was based largely on the presence of Cape Channel and the island's low and narrow profile in 1961. He showed that this inlet was recorded as open on 11 maps between 1585 and 1657. On four of those maps (1585, 1590, 1606, 1657), which were analyzed in this study, however, evidence of an open inlet was lacking. During the 1962 (Ash Wednesday) storm, an ephemeral inlet was opened at this site; because of the conflicting evidence it was not included in Figure 9 as a persistent inlet.

Accuracy of inlet location

29. Use of past and present inlets in an attempt to develop a relationship between inlets and shoreline change requires that historic, and if possible, prehistoric inlets be identified and accurately located. NOS shoreline maps were used to compare inlet locations after 1852 with those on maps used in Cumming's (1966) report. The locations of inlets on the NOS maps are considered accurate.

30. In Figure 9, the two positions listed for Oregon Inlet in 1861 (Bachman and Colton in Cumming 1966) are included to show the possible variation due to (a) cartographic mislocation of the inlets or (b) mislocation during the original survey work. The Bachman map is entitled "Panarama [sic] of the Seat of War, Birds Eye View of North and South Carolina, a part of Georgia." It is an oblique map, very schematically drawn, lacking latitude and longitude coordinates, and the location and existence (or nonexistence) of particular inlets on it should be viewed with caution. The Colton map of the same year is more accurately drawn, though its description emphasized the railroad and overland transportation routes and no discussion of the coastline is included. There is an obvious lack of correlation between the general trend of Oregon Inlet's present position and the 1861 positions; however, if the entire sequence of inlets shown on the Colton Map (i.e., Oregon, New, unnamed, and Loggerhead) are shifted to the north by 3'-5' of latitude, this dramatic offset is eliminated. This seems to be a more reasonable solution than keeping the inlets in their 1861-mapped position and assuming a "zig-zag" migration pattern.

25

31. Navigation accuracy increased through time. Evidence of this grow-ing sophistication in positioning tools and techniques can be seen in the noticeable decrease in the width of the inlet sequences. Prior to the 19th century, New Currituck, Roanoke, and Chickinacommock Inlet sequences are all at least 5' of latitude wide, while the sequences of Oregon (assuming that the offset of 1861 is a distortion), New, and New/Loggerhead Inlets are all approximately 3' of latitude wide. With time, more accurate measurements re-sulted is less lateral variation in the map position of inlets. Inlet loca-tions on maps other than those produced by NOS are potentially inaccurate by up to ±5 minutes of latitude (i.e., the approximate amount by which early measurement methods could vary and still produce a map with the general con-figuration of the existing shoreline). This reasoning could help to explain the dramatic change in location of Roanoke Inlet between 1657 and 1770 shown in Figure 9.

32. Another explanation for the variation over time of inlet locations is north or south inlet migration. The NOS maps show, for example, that Oregon Inlet has migrated south over the past 130 years, at the rate of 29 m/ year, for almost 2 minutes of latitude. If the migration sweep of other inlets falls within the same range, an inlet remaining open for 100 years or so could move ±2 minutes in latitude.

33. A large shoreline bulge into the sound is often a good geomorphic clue to the presence of an inlet which was open in the past. Currituck Inlet is clearly related to a wide section of the island (Figure 11). Musketo Inlet, however appears to be located several minutes north of a large island bulge (Figure 9); quite likely, the bulge is the site of historic Musketo Inlet. The very large width change from Kitty Hawk to Nags Head is likely re-lated to prehistoric Kitty Hawk Inlet (Figure 12). Roanoke Inlet, shown as having varied widely in an alongshore distance (Figure 9), is centered at a shoreline bulge; possibly more than one inlet existed in this reach. An island bulge is also associated with the sites of the now-closed New and Loggerhead Inlets shown on NOS shoreline maps (Figure 11).

Continental Shelf

34. A barrier island shoreline is to a large extent shaped by ocean waves which move across the continental shelf onto the shoreface and break

26

Figure 12. Probable site of a large pre-1585 inlet at Kitty Hawk,
N. C. (Inlet was located in the left-central part of the picture
(Fisher 1962). Note the longshore bar as evidenced by breaking
waves in the Atlantic Ocean at the right side of the photograph.
The Wright Brothers Memorial is located in the left foreground)

near the beach. The shoreface, or inner part of the continental shelf, has a
concave profile; seaward of the shoreface, the continental shelf is planar
and dips away from the coast. Because wave form is modified and wave energy
is dissipated in shallow water, the width of the continental shelf is a factor
in regulating the amount of wave energy which reaches and is expended at the
coast. The shelf width to the 180-m (100-fathom) isobath narrows from 126 km
east of Cape Henry to 48 km east of Cape Hatteras.

35. Because of its decreasing width, the slope of the continental shelf
increases from north to south. The depth at the base of the steep, concave
shoreface also increases in that direction (Figure 13). The profiles shown
in the figure are averages of nine profiles, spaced 1.5 km apart, at each
location. An analysis of seismic data from the study area suggests the shoreface may be resting unconformably upon older sediments of the planar and
seaward-dipping continental shelf; this implies the concave shoreface is
shaped by processes (waves, currents) active today or in the recent past.

27

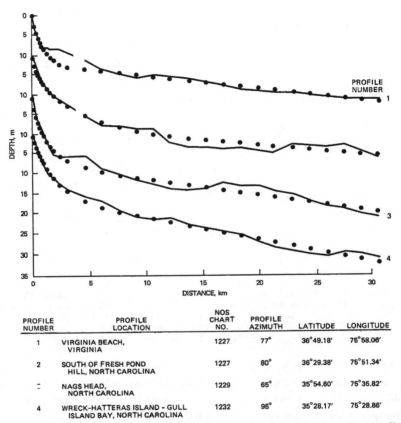

PROFILE NUMBER	PROFILE LOCATION	NOS CHART NO.	PROFILE AZIMUTH	LATITUDE	LONGITUDE
1	VIRGINIA BEACH, VIRGINIA	1227	77°	36°49.18'	75°58.06'
2	SOUTH OF FRESH POND HILL, NORTH CAROLINA	1227	80°	36°29.38'	75°51.34'
3	NAGS HEAD, NORTH CAROLINA	1229	65°	35°54.60'	75°35.82'
4	WRECK-HATTERAS ISLAND - GULL ISLAND BAY, NORTH CAROLINA	1232	95°	35°28.17'	75°28.86'

Figure 13. Continental shelf profiles taken between Virginia Beach, Va., and Hatteras Island, N. C., to 30 km from shore (the averaged profile is a solid line; dotted profile is a mathematical fit to the average profile) (after Everts 1976)

36. Bathymety further seaward on the continental shelf reflects both past and present processes. In the study area, the shelf is a broad sand plain molded into north-south-trending sand ridges and troughs of up to 10-m relief (Swift et al. 1978a, p 21). Two shelf-valley complexes were generated by the landward displacement of the Chesapeake Bay and Albemarle Sound estuaries as sea level rose in the Holocene epoch. Each complex has left an imprint on the inner shelf: one lies seaward of the shoreface of Virginia Beach, Virginia, and the other off Nags Head, North Carolina (Swift et al. 1978a, p 20).

28

37.   Four clusters of closely spaced ridges trend oblique to the shoreline and tie to the shoreface between Cape Henry and Cape Hatteras. They are,
from north to south, False Cape Shoals, Oregon Shoals, Wimble Shoals, and
Kinakeet Shoals. Swift et al. (1978b, p 270-271) note these characteristics
of shoreface-connected ridges: (a) the ridges rest on surfaces exposed as
the shoreface retreats to the west, (b) shoreface-connected ridges form angles
with the coast opening into the direction of prevailing flow (i.e., from north
to south), (c) sand on the seaward (downcurrent) flanks is finer than sand on
the landward (upcurrent) flanks, (d) ridges tend to be steeper on the seaward
side except next to their shoreface connection, and (e) ridges tend to migrate
downcoast and offshore. These ridges are emphasized because they appear to
have an influence on adjacent shoreline retreat rates.

Tides, Winds, and Waves

38.   The data in paragraphs 39-46 are presented for reference only and
are not used in the analysis section; they provide background on the dynamic
conditions which have existed in recent times.

Tides and other sea level fluctuations

39.   An astronomical tide is the periodic rising and falling of the
water surface resulting from the gravitational attraction of the moon and sun
on the rotating earth. The period of a complete tidal cycle in the study
area is 12.4 hours; the mean and spring tide ranges from NOS Tide Tables for
1981 are shown in Figure 14.

40.   Sometimes superimposed on the astronomical tides is storm surge;
i.e., wind and wave setup and water surface differences caused by barometric
variations. Wind setup is the vertical rise in water level at a lee shore
caused by wind shear stresses on the water surface. Wave setup is another
superelevation of the water surface, caused by the onshore mass transport of
water by waves. In the sounds 20 km or more away from Oregon Inlet, the
astronomical tide range is less than 0.3 m, but the wind setup may raise the
water surface a meter or more for a 1-year wind event. Return periods for
storm surge along the ocean shore at Kitty Hawk are shown in Figure 15 (Ho
and Tracey 1975).

41.   A changing sea level occurring over a period of years may have a
profound effect on shoreline position. Changes on the order of years in sea

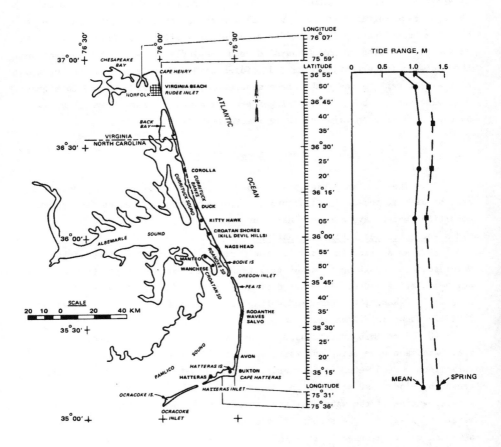

Figure 14. Tide ranges, Cape Henry to Cape Hatteras (from NOS tide
tables, 1981, for ocean sites (above mlw))

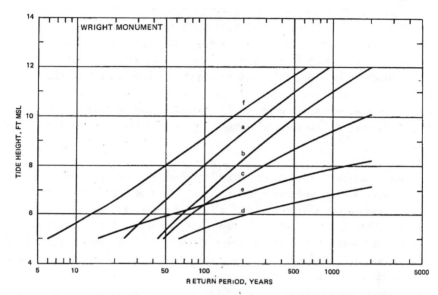

Figure 15. Tide frequencies for the ocean shoreline at Kitty Hawk, N. C., for several classes of storms: (a) landfalling, (b) alongshore, (c) inland, (d) exiting hurricanes and tropical storms, (e) winter storms, (f) all storms (from Ho and Tracey 1975)

surface elevation may cause a reshaping of the beach, nearshore, and inner continental shelf profile; a rising sea relative to land will probably cause the shoreline to retreat.

42. Sea level change data are not available in the study area. However, tide gage records (Hicks 1981) from Norfolk, Virginia, and Charleston, South Carolina, exist, respectively, for the periods 1928 through 1978 and 1922 through 1978. The average rate of sea level rise relative to land at Norfolk was +4.4 mm/year, but the trend may be one of a declining rise rate (Everts 1981); from 1940 to 1978 the average was +3.7 mm/year, or about 15 percent less than the 1928 to 1978 average. At Charleston the 1922-to-1978 average was +3.6 mm/year, but Hicks' (1981) data show a 1940-to-1978 rate of only +2.5 mm/year and indicate a decline in the rate of sea level rise relative to land.

Wind conditions

43. Wind direction and mean scalar speed in the study area are given in Figures 16 and 17. Mean annual velocities increase slightly to the south. A velocity of 16 km/hour, 10 m above the ground, is required to initiate sand movement. Speeds of 25 km/hour are required to sustain transport (Bagnold 1941). Winds at or above these speeds are predominantly onshore from the northeast and occur most frequently during the winter months. The effects of northwest winds, which are potentially important, may be lessened because of local sheltering due to forests on the west side of the barriers (Hennigar 1979).

Waves

44. Changes in shoreline configuration result from a combination of (a) wave action which mobilizes sediment and (b) wave-, wind-, and tide-induced currents which transport the mobilized sediment.

45. Wave data are available from gages situated at Virginia Beach, Virginia, and Nags Head, North Carolina (Thompson 1977). The Virginia Beach gage, located at a depth of 5.5 to 6 m of water msl on the north side and near the seaward end of the 15th Street fishing pier, was a step resistance, staff relay gage in noncontinuous operation between 1962 and 1971. At Nags Head a step resistance, staff relay gage was in operation, with some short periods of inoperation, between 1963 and 1972. In 1972 a continuous wire staff gage was installed at a depth of 5 m of water msl on the north side and 50 m from the end of Jeannettes Fishing Pier. A third gaging site, recently operational, is the CERC Field Research Facility, just north of Duck, North Carolina. Wave data have been available from that site since 1979.

46. The Wave Information Study, Phase III (Jensen 1983), provides hindcast wave data for 20-year time periods for the study area. Using those hindcast results, Figure 18 shows the annual cumulative significant wave height distribution for waves which approach from all directions at station 81 at a 10-m water depth off Kitty Hawk, North Carolina. The mean and maximum significant wave heights are, respectively, 0.89 m and 4.70 m. Figure 19 is a wave rose diagram for the same location off Kitty Hawk showing the significant wave height and direction of wave propagation for the combined 20-year hindcast data.

32

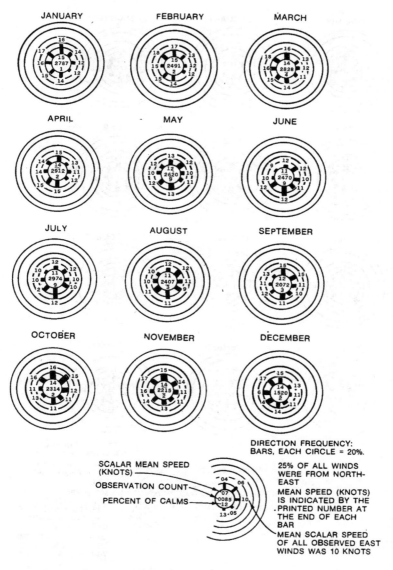

Figure 16. Surface wind rosés, Cape Henry and vicinity, from data collected 1850-1960

33

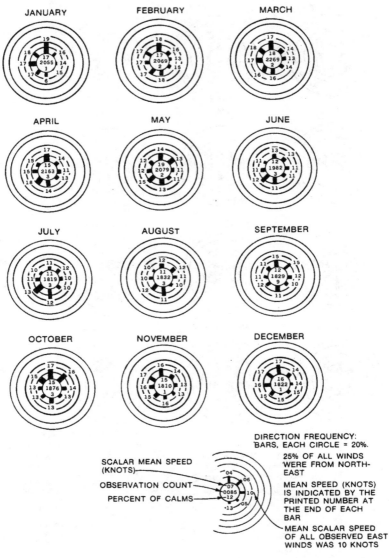

Figure 17. Surface wind roses, Cape Hatteras and vicinity, from data collected 1850-1960

34

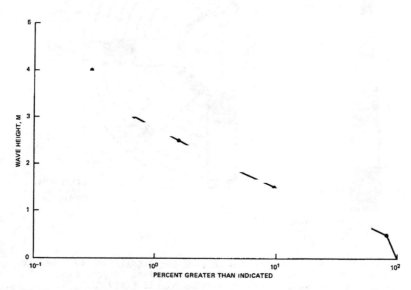

Figure 18. Annual cumulative significant wave height distribution based on
20 years of hindcast data measured at 10-m water depth off Kitty Hawk, N. C.
(after Jensen, draft report)

35

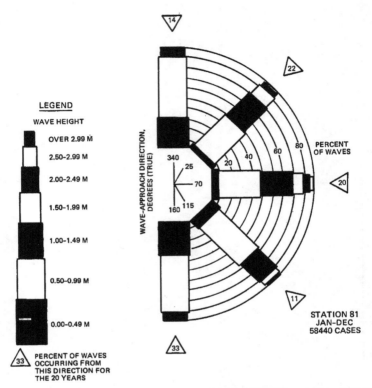

Figure 19. Wave rose diagram showing the significant wave height
and direction of wave propagation for combined 20-year hindcast
data in a 10-m water depth at station 81 off Kitty Hawk, N. C.
(from Jensen 1983)

Coastal Storms

47. Extratropical (northeasters) and tropical (hurricanes) storms play
a major role in changing the position of the shoreline. Figure 20 from Ebersole (1982) shows the frequency of occurrence of storm surge for extratropical
storms at Hampton Roads, Virginia, the closest tidal reference station to the
study area (about 20 km west of Cape Henry). The figure was produced using
hourly values of water level data from 1952 to 1971. Ebersole (1982) found
that about 20 years of data provided a relatively stationary tidal probability

36

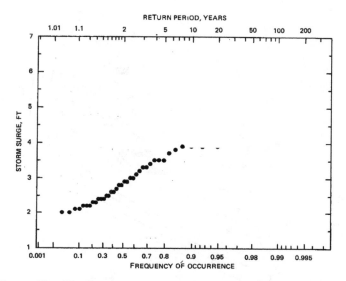

Figure 20. Yearly storm surge return period for extratropical
storms at Hampton Roads, Va. (after Ebersole 1982)

density function. Figure 20 illustrates the yearly return period for extratropical storms which produce the given storm surge elevations.

48. Hayden (1975) studied secular variations in storm occurrence. In a
hindcast study of extratropical storms (i.e., with waves greater than 1.6 m),
he found that storm occurrences were at a maximum in March between 1942 and
1960, but that the maximum had moved to January by 1974. The mean annual
number of storms with waves over 1.6 m did not vary significantly (the annual
average was about 34); however, Hayden (1975, p 982) found the number of
events in which the waves exceeded 2.5 m had increased 1.9 times between the
1942-1965 period and the 1965-1974 period. Hayden notes that the increased
frequency of large stormwaves is consistent with observed trends in shoreline
erosion.

49. Hurricanes generally move from southwest to northeast in the study
area (Ho and Tracey 1975), with an increase in the frequency of hurricanes
from Cape Henry to Cape Hatteras (Figure 21). Simpson and Riehl (1981, p 109,
292) show below-average frequencies predominated from 1895 to 1930; in 1931

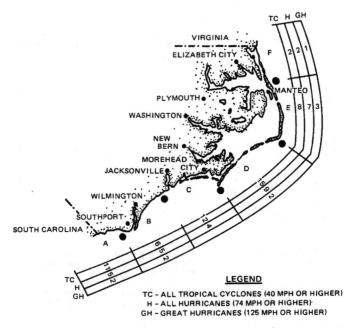

Figure 21. Number of tropical cyclones (hurricanes) reaching the North Carolina coast, by sector, for the period 1886-1970

hurricane frequency rose to an above average level and remained there until 1960 when another decline began.

Coastal Structures

50. Various types of structures have been constructed on the beach and in the nearshore zone along the study area coast (Table 3). They were constructed to serve four general needs: recreation and research (piers, Figure 22), coastal protection (bulkheads), coastal stabilization (groins, Figure 23), and navigation improvements (jetties, Figure 24). Jetties and groins modify the directional distribution of energy approaching the adjacent shoreline and act as barriers to longshore sand transport; piers also modify the movement of sand in an alongshore direction.

51. In addition to fixed coastal structures, beach fills and dunes or

38

Figure 22. Recreational pier at Virginia Beach, Va., a typical
fishing pier for the study area (note the shoreline bulge
created at the pier by a decrease in the rate of sediment
transport parallel to shore. The decrease probably occurred
because the pier piles created a partial obstruction to shore-
parallel flow and because the pier piles in the water dissipate
some of the wave energy)

Figure 23. View toward north of groins near Cape Hatteras
Lighthouse; Cape Hatteras Point is about 1 km south of the
lighthouse (note the change in shoreline orientation at
the groin system. Sand moving from north to south is
partially trapped and held north of the groins. The beach
to the south (bottom of photograph) consequently receives
less sand than it loses, and erosion is accelerated)

40

Table 3

Coastál Structures, Cape Henry to Cape Hatteras*

Location	Structure	Number	Remarks
West of Lynnhaven Inlet, Va.	Fishing pier	1	
Virginia Beach, Va.	Fishing pier	1	
Virginia Beach, Va.	Bulkhead		
Virginia Beach, Va.	Jetties	2	Rudee Inlet, South Structure, is a weir jetty
Duck, N. C.	Research pier	1	550 m long or 2 to 3 times as long as the fishing piers located south of Kitty Hawk, N. C.
Kitty Hawk to South Nags Head, N. C.	Fishing pier	5	
Rodanthe, N. C.	Fishing pier	1	
Salvo, N. C.	Fishing pier	1	
Cape Hatteras Light, N. C.	Groins	3	

* Oceanfront only, 1980 conditions.

Figure 24. Weir jetty system at Rudee Inlet, Virginia Beach, Va.
(Sand movement at this location is predominantly south to north
(left to right in the photograph). Sand moves over the low weir
section and is periodically pumped to nourish the recreation
beaches north of the north jetty)

sand fences have been constructed to prevent flooding from the Atlantic Ocean
and to slow or halt shoreline retreat. The most extensive beach fill efforts
have taken place at Virginia Beach, where sand has been placed on the beach
for the last 25 years: between 1952 and 1976, over 4.5 million cu m of sand
were placed along 8 km of shoreline, mostly within the 5.5-km reach north of
Rudee Inlet (Goldsmith et al. 1977). Sand sources were (a) a stockpile at
Cape Henry where material dredged from Thimble Shoal Channel in the Chesapeake
Bay entrance was stored, (b) Lakes Rudee and Wesley and Owl Creek, (c) Lynn-
haven Waterway, and (d) upland borrow sources (currently from south of Rudee
Inlet). The net alongshore movement is about 200,000 cu m of sediment/year to
the north at Rudee Inlet. Bypassing is presently accomplished using the weir
jetty system shown in Figure 24: sand passes over the low weir crest into a
sheltered depositional basin from which it is periodically pumped north across
Rudee Inlet to the Virginia Beach problem area.

    52. In response to rapid shoreline retreat north of Cape Hatteras, the

National Park Service contracted to have 240,000 cu m of sand placed on the beach in 1966 (Dolan 1972a). That fine-grained sand, taken from Pamilico Sound, was soon lost. Three groins were then constructed by the U. S. Navy in 1970. Further erosion north of the groins was addressed by a beach-replenishment project in 1972 when 170,000 cu m of beach sand from Cape Point was placed; in 1973, 750,000 cu m were added from the same source.

53. Between 1936 and 1940, sand fences were built along various reaches of the study area by the Civilian Conservation Corps to create and maintain continuous dunes (Dolan 1972b). Over 900,000 m of fencing was erected on Bodie, Pea, and Hatteras Islands, most of it near the beach. Following a severe storm in March 1962, a dune was constructed along 30 km of oceanfront between Nags Head and Kitty Hawk. In the 1950's and 1960's the U. S. Department of the Interior, National Park Service, constructed and stabilized dunes in the Cape Hatteras National Seashore (i.e., from South Nags Head to past the southern limit of the study area).

Data Sources

54. Forty-two historical NOS and C&GS shoreline surveys and maps, at scales varying from 1:5000 to 1:40,000 and dating from 1849 through 1975, exist for the study area. The earliest surveys (up to around 1927) were "topographic surveys" and were practically all completed by planetable. Since 1927, aerial photography and photogrammetric methods (thus photogrammetric surveys) have been used increasingly to provide topographic information along the coast (Shalowitz 1964, p 52).

55. Eighteen 1:24,000-scale U. S. Geological Survey (USGS) quadrangles were selected to be the base maps for this project (Figure 25). They were revised by the Cartographic Revision Section of the Photogrammetry Division of NOS with 1:24,000-scale color photography, taken on 16 March 1980 at near high water, covering both sides of the barrier island and all of the ocean coast within the project. This procedure is described more fully below. The historical sheets available for each base map, with their scales and dates of survey, are listed in Table 4. A particular sheet may often be listed on more than one base map; each base map usually comprises sheets of varying scales and area limits.

Shoreline Definition

56. Topographic surveys, in support of hydrographic surveys, have been compiled by NOS since the early 1800's. These surveys are the basis for the delineaton of the shoreline on the nautical charts published by the Agency. According to Shalowitz (1964), the authority on the historical significance of early topographic surveys of NOS, "The most important feature on a topographic survey is the high-water line." High-water line (HWL) is a general term; because it is used in this report as the shoreline, it must be defined as actually surveyed through the years by NOS and its predecessors.

57. About 1840, Ferdinand Hassler, the first Superintendent of the Survey, issued the earliest instructions for topographic work. Those instructions (Volume 17, Coast Survey, Scientific, 1844-1846, handwritten) included the following:

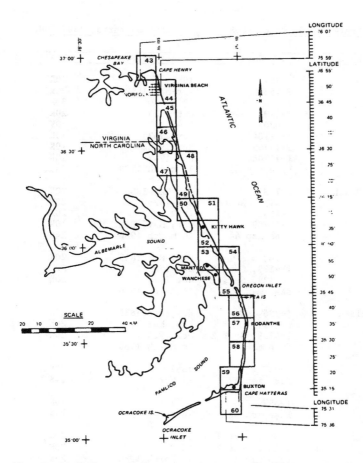

Figure 25. U. S. Geological Survey 1:24,000 quadrangles used as base maps in this study

17. On the sea shore and the rivers subject to the tides, the high and low water lines are to be surveyed accurately; and the kind of ground contained between them, whether sand, rock, shingle or mud marked accordingly. The low water line is taken by offsets whilst running the high water, and when not too far apart from each other, but when their distance is great they must be surveyed separately: a couple of hours before the end of the ebb, and the same time

45

Table 4·

Historic Shoreline Surveys, Cape Henry to Cape Hatteras, 1847-1980

Sheet Name*	T-Sheet	Date of Survey	Scale	Map Number
Cape Henry	T-507	1852	20K	1
	T-753	Apr-May 1859	20K	2
	T-3647	1916	30K	3
	T-4139, Sect 1	Oct 1925	20K	4
	T-8301	1944	20K	5
	T-11704	May 1962	10K	6
	T-11705	May 1962	10K	7
	T-11706	May 1962	10K	8
	Base map 43	16 March 1980	24K	43
Virginia Beach	T-753	Apr-May 1859	20K	2
	T-4139, Sect 1	Oct 1925	20K	4
	T-8299	1942	20K	9
	T-11709	Mar, May, Sept 1962	10K	10
	Base map 44	16 March 1980	24K	44
North Bay	T-743	Feb, Mar 1859	20K	11
	T-4139, Sect 2	Oct 1925	20K	4
	T-4139, Sect 1	Oct 1925	20K	4
	Base map 45	16 March 1980	24K	45
Knotts Island	T-736	Nov, Dec 1858	20K	12
	T-743	Feb, Mar 1859	20K	11
	T-4139, Sect 2	Oct 1925	20K	4
	Base map 46	16 March 1980	24K	46
Barco NW	T-657	1857	20K	13
	Base map 47	16 March 1980	24K	47
Barco NE	T-657	1857	20K	13
	Base map 48	16 March 1980	24K	48
Barco SE	T-381, Sect 1	1852	20K	14
	T-381, Sect 2	1852	20K	14
	Base map 49	16 March 1980	24K	49
Powells Point NE	T-381, Sect 2	1852	20K	14
	Base map 50	16 March 1980	24K	50
Kitty Hawk NW	T-292	1849	20K	15
	Base map 51	16 March 1980	24K	51

(Continued)

* U. S. Geological Survey quadrangle; see Figure 25.

(Sheet 1 of 3)

Table 4 (Continued)

Sheet Name	T-Sheet	Date of Survey	Scale	Map Number
Kitty Hawk SW	T-292	1849	20K	15
	T-3538	1915	40K	20
	Base map 52	16 March 1980	24K	52
Manteo	T-351	1851	20K	16
	T-3538	1915	40K	20
	T-9159	Dec 1949	20K	17
	Base map 53	16 March 1980	24K	53
Roanoke Island NE	T-354	Jan 1849	20K	18
	T-3538	1915	40K	20
	T-9160	May 1949	20K	19
	Base map 54	16 March 1980	24K	54
Oregon Inlet	T-354	Jan 1849	20K	18
	T-3538	1915	40K	20
	T-9278	Dec 1949	20K	21
	T-11672	1963-64	10K	22
	T-11665	1963-64	10K	23
	T-12140	1963-64	10K	24
	TP-00887	1975	5K	25
	TP-00889	1975	5K	26
	Base map 55	16 March 1980	24K	55
Pea Island	T-367	Mar, Apr 1852	20K	27
	T-3707	1917	40K	28
	T-8711, Sect 1	1946	10K	29
	T-8711, Sect 2	1946	10K	29
	T-12147	May 1962	10K	30
	T-12562	Oct 1963	10K	31
	Base map 56	16 March 1980	24K	56
Rodanthe	T-367	Mar, Apr 1852	20K	27
	T-3707	1917	40K	28
	T-8712, Sect 1	1946	10K	32
	T-8712, Sect 2	1947	10K	32
	T-12437	April 1963	20K	33
	Base map 57	16 March 1980	24K	57
Little Kinnakeet	T-377	Jan, Feb 1852	20K	34
	T-3707	1917	40K	28
	T-8713, Sect 1	1946	10K	35
	T-8713, Sect 2	1946	10K	35
	T-12438	April 1963	20K	36
	Base map 58	16 March 1980	24K	58

(Continued)

(Sheet 2 of 3)

Table 4 (Concluded)

Sheet Name	T-Sheet	Date of Survey	Scale	Map Number
Buxton	T-377	Jan, Feb 1852	20K	34
	T-790	1860	20K	37
	T-1246	1872	20K	38
	T-3707	1917	40K	28
	T-8714, Sect 1	1946	10K	39
	T-8714, Sect 2	1946	10K	39
	TP-00507	April 1974	20K	40
	Base map 59	16 March 1980	24K	59
Cape Hatteras	T-377	Jan, Feb 1852	20K	34
	T-790	1860	20K	37
	T-1246	1872	20K	38
	T-3707	1917	40K	28
	T-8718	1947	10K	41
	T-12442	April 1963	20K	42
	TP-00507	April 1974	20K	40
	Base map 60	16 March 1980	24K	60

during the commencement of the flood tides will be the
proper time for taking the low water line, and your
operations must be so timed, as to be on the shore on
those periods.

18. You will establish points along the shores, and
mark them securely by means of stakes, at suitable
distances, for the use of the hydrographical parties
in taking their sounding--and also furnish them with
the high and low water line, from your map, they may
require.

58. The first specific instruction regarding the nature of the line to
be surveyed is contained in the Plane Table Manual (Wainwright 1889), which
states: "In tracing the shoreline on an exposed sandy coast, care should be
taken to discriminate [sic] between the average high-water line and the storm
water line." Still later, Shalowitz (1964, p 174) elaborated by stating:

The mean high-water line along a coast is the intersection of the plane of mean high water with the shore.
This line, particularly along gently sloping beaches,
can only be determined with precision by running spirit
levels along the coast. Obviously, for charting purposes, such precise methods would not be justified,
hence, the line is determined more from the physical
appearance of the beach. What the topographer actually delineates are the markings left on the beach by
the last preceeding high water, barring the drift cast
up by storm tides. On the Atlantic coast, only one
line of drift would be in evidence....If only one line
of drift exists, as when a higher tide follows a lower
one, the markings left by the lower tide would be
obliterated by the higher tide and the tendency would
be to delineate the line left by the latter, or possibly a line slightly seaward of such drift line.

In addition to the above, the topographer, who is an
expert in his field, familiarizes himself with the tide
in the area, and notes the characteristics of the beach
as to the relative compactness of the sand (the sand
back of the high-water line is usually less compact
and coarser), the difference in character and color
of the sun cracks on mud flats, the discoloration of
the grass on marshy areas, and the tufts of grass or
other vegetation likely along the high-water line.

59. Historical references are included to emphasize that it was the
intention of all the agency's topographic surveys to determine the line of
mean high water (MHWL) for delineation on maps. With the exception of tidal
marsh areas, where in most cases the outer limit of vegetation is mapped,
the MHWL delineated on the surveys by the experienced topographer or

49

photogrammetrist was that line at the time of the survey or the date of photography.

60. With the advent of precision aerial photography, the compilation of a "T-sheet" for photohydro support opened a new dimension in shoreline compilation. When stereoscopic instruments and known tide data were used, the MHWL could be accurately determined by aerial photography. This method was supplemented, when possible, by profile points run from vertical bench marks to verify the photointerpretation. When beach profiles were run on the more contemporary surveys, they were referenced to the nearest tide station. If this was a tertiary (i.e., temporary) station, the readings were referenced to a primary station.

Methods Used to Revise the 1980 Mean High Water Line

61. To make this study as current as possible, USGS quadrangle maps were revised to show a 1980 MHWL. The revision was made using 1980 color aerial photographs flown for this study. Date and time of the photography were correlated with the stage of the tide, and a detailed stereoscopic examination of the photographs was made to determine the 1980 MHWL. This process was completed by the Cartographic Revision Section of the Photogrammetry Division of NOS. Their method was by direct transfer of the photointerpreted line (see paragraph 60) from 1:24,000-ratioed film positives to the USGS base maps. Using the ratioed photography, the base maps (manuscripts) were held planimetrically to local physical features. In absence of triangulation stations to position the manuscript accurately against the photographs, it is possible to use "hard" planimetric features, such as road intersections or other permanent physical structures without great relief, to assure good photographic positioning. In areas where there were not enough features to assure proper positioning, stereomodels were set on the National Ocean Survey Analytical Plotter (NOSAP). NOSAP is a high-precision stereoscopic plotter that allows the operator to bridge over areas of sparse control and accurately determine the correct relationship between photographic models and the base maps. Due to time restraints, no field check of the office-determined 1980 MHWL was made. All shorelines compiled by this method were reviewed to assure a uniformity of the photointerpreted shoreline, accuracy of compilation, and

proper symbolization. These maps were then digitized, checked, and reviewed in the manner identical to that used for all historical source maps.

Data Reduction Procedures

62. Copies of all historical maps used as source data in this study were obtained from the NOS vault in Riverdale, Maryland, through the NOS Reproduction Division. Copies were initially bromide prints (a photographic process which provides a long shelf-life copy) and were later made into more stable matte-finish film positives. Historical sheets covering the study area were examined to determine which sections of shoreline would be included in the study, and those were highlighted using a yellow felt marker. Only those areas and sections of shoreline for which data from other NOS historical maps would be available for comparison were used.

63. Digitizing of the shoreline on each historical map and each base map, revised for contemporary shoreline, was the next task. This procedure was completed by the Data Translation Branch, Environmental Data and Information Service, Asheville, North Carolina. The digitizing was completed on a Calma-graphics III system, with a repeatability factor of ±0.001 in. and a maximum absolute error of ±0.003 in. The digitized data tapes were then processed using a program developed by the NOS Marine Data Systems Project for use with the NOAA UNIVAC computer (GPOLYT2); this program allows for the conversions of the digitized data to geographic positions (GP's). Since many of the historic sheets used in the study were completed before the North American Horizontal Datum of 1927 (NA 1927) was established, the GP's for these sheets were converted to that datum so that accurate comparisons between pre- and post-NA 1927 surveys could be made. Conversion was completed mathematically, based on the conversion factors for triangulation stations in the area, on a program also written by the NOS Marine Data Systems Project.

64. After processing of the data was completed, plot tapes were generated using the NOS McGraphics program, and the plot tapes were used, with a Calcomp 748 plotter and Calcomp 925 Controller, to plot the shoreline movement maps. This task was completed with the assistance of the NOS Automated Cartography Group.

65. All sections of shoreline from the source maps were digitized so that all shoreline points could be converted into GP's and replotted at any desired scale (before the final portrayal scale of 1:24,000 for the shoreline movement maps was chosen, other scales were tested to determine which map scale would portray the data in the most readable form). Digitizing also removed inherent media distortion caused by the age of the original manuscripts. The mechanics and mathematics of the digitizing system required that all projection (latitude and longitude) intersections completely enclose the data to be digitized. By assigning known and true values for each projection intersection, the GPOLYT2 program adjusted each of the shoreline points enclosed within a projection cell, based on the true values of the intersections versus the digitized and computed values for those same intersections. The values for each shoreline point are thus correct in their position relative to the known (true) projection intersections and to known triangulation data (Figure 26).

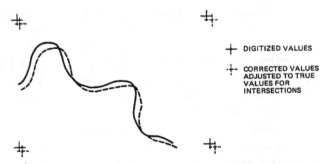

DIGITIZED VALUES

CORRECTED VALUES
ADJUSTED TO TRUE
VALUES FOR
INTERSECTIONS

Figure 26. Digitization procedure for correcting shoreline position locations when original shoreline movement map distortions exist

66. Following the digitizing process, each sheet was reviewed visually with the use of a raw data plot in which shoreline positions were shown at the same scale as the original map. The plotted shoreline was superimposed on the original map and checked for completeness and accuracy of tracking during digitization. This review helped to minimize a potential source of human error that could occur during the digitizing process.

67. Other sources of potential error also were considered. The most difficult of these to determine accurately was the location accuracy of the MHWL on the source surveys and maps, on either (a) the early surveys prior to approximately 1930 and (b) the group of maps based on photogrammetric surveys. In discussing the early surveys, Shalowitz (1964, p 175) has stated:

> The accuracy of the surveyed line here considered is that resulting from the methods used in locating the line at the time of survey. It is difficult to make any absolute estimates as to the accuracy of the early topographic surveys of the Bureau. In general, the officers who executed these surveys used extreme care in their work. The accuracy was of course limited by the amount of control that was available in the area.

> With the methods used, and assuming the normal control, it was possible to measure distances with an accuracy of 1 meter (Annual Report, U. S. Coast and Geodetic Survey 192 (1880)) while the position of the plane-table could be determined within 2 or 3 meters of its true position. To this must be added the error due to the identification of the actual mean high water line on the ground, which may approximate 3 to 4 meters. It may therefore be assumed that the accuracy of location of the high-water line on the early surveys is within a maximum error of 10 meters and may possibly be much more accurate than this. This is the accuracy of the actual rodded points along the shore and does not include errors resulting from sketching between points. The latter may, in some cases, amount to as much as 10 meters, particularly where small indentations are not visible to the topographer at the planetable.

> The accuracy of the high-water line on early topographic surveys of the Bureau was thus dependent upon a combination of factors, in addition to the personal equation of the individual topographer. But no large errors were allowed to accumulate. By means of the triangulation control, a constant check was kept on the overall accuracy of the work.

On aerial photographs, the MHW line is located to within 0.5 mm at map scale (USC&GS 1944). This translates to less than 5 m on the ground for a map scale of 1:10,000, or 9.99 m on the ground for a map scale of 1:20,000. Since the great majority of source maps were of a larger scale than the 1:24,000 base maps, the 0.5-mm accuracy of source maps made using aerial photography was at least maintained by reducing most of the source maps to the common base scale of 1:24,000. Present NOS survey maps are even more accurate. In a recent shoreline mapping project in the state of Florida using NOS charts, 36 random features such as road intersections and shoreline features, including points of marsh, were scaled from the map compiled from aerial photography. Where these features were then located by field traverse and the geodetic coordinate values compared, the check revealed a maximum error of ±3.0 m. This accuracy is not claimed for all surveys, but it does serve as an indicator of the accuracy of surveys conducted within NOS.

68. The last source of potential error is the conversion of digitized values to GP's. Digitizing equipment automatically recorded 1,000 coordinate values for every inch of shoreline traced, which values were then corrected to true latitude and longitude positions, as previously discussed. The GPOLYT2 program printout provided a final error column each for "Latitude Y" and "Longitude X," which were examined on each printout. In the event any of the figures exceeded 0.5 mm (at map scale), the digitizing effort was rejected and the original sheet was redigitized. Although the maximum allowable error from this source was 4.99 m on the ground for a 1:10,000-scale map and 9.99 m on the ground for a 1:20,000-scale map, rarely were the error column values as high as 0.5 mm; in most cases, they were 0.2 mm or smaller. As such, the possible errors from this source were more likely to be in the vicinity of 1.99 m on the ground for a 1:10,000-scale map and 3.99 m on the ground for a 1:20,000-scale map. Since most data were finally portrayed at a scale smaller than the map being digitized, the shoreline movement maps produced are well within map accuracy standards.

PART IV: DATA ANALYSIS AND DISCUSSION

69. Reproductions of composite shoreline movement maps are enclosed separately. These maps are useful in a qualitative way; i.e., they provide an easy means of observing the changes that have occurred in the past. Because of slight variations in shoreline position created in the printing process, however, they should not be digitized for quantitative use in coastal management, engineering, or research. To enable the data these maps represent to be so used, the following paragraphs describe the techniques used in this study to quantify shoreline change.

70. An analysis routine was used to average shoreline change parameters for specified longshore distances. Because geographic point analyses were based on latitude and longitude, a reasonable distance to use was one keyed to those measures. Based on shore orientation, a 1-minute-latitude (about 2 km) or -longitude (about 1.5 km) distance was selected to average long-term shoreline changes. It deserves mention that the shoreline change rate given is the average for the entire shoreline within the 1-minute coastal reach, not the rate at particular sites 1 minute apart. The distinction is an important one because measurements made at a constant alongshore interval seem to be subject to a bias depending upon the particular interval chosen (Hayden et al. 1979).

Analysis Methodology

71. Shoreline change rates resulting from the following analyses, although averages in space, are based on particular points in time. Nothing is included that identifies what happened to the shoreline in the interval between shoreline surveys; the analyses simply distribute the change uniformly over the separating time increment. The rates given in this report are the shore-normal rates of movement averaged for a fixed shore-length increment; they were obtained using changes in plan area between successive latitude or longitude boundaries 1 minute apart. For each survey set from time t a plan area A(t) was specified using fixed latitude and longitude boundaries (three of the boundaries used) and the shoreline (the fourth boundary) (Figure 27). The latitude and longitude boundaries were invariant in time; only the shoreline boundary changed. That change in shoreline position between surveys

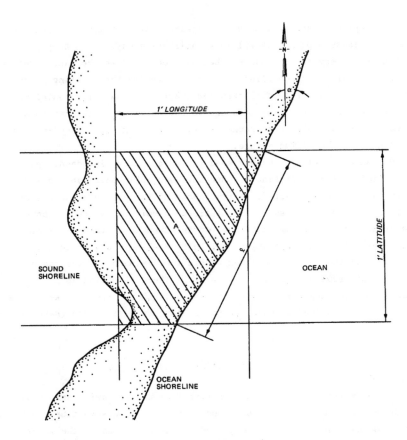

Figure 27. Definition sketch illustrating parameters used to obtain
shoreline change rates for a north-south-trending ocean shoreline
(a north-south trend is defined as $315° \leq \alpha \geq 45°$)

created a change in plan area. The difference in plan area for each time
interval, divided by the shoreline length ℓ and the number of years between
surveys, produced an annual shoreline change rate S_i for a particular survey
interval

$$S_i = \frac{A(t_i) - A(t_{i-1})}{(t_i - t_{i-1})\ell} \tag{1}$$

where i varies from 2 to n, and n equals number of surveys. This shore-line change rate is the average shore-normal movement landward (-) or seaward (+) of the shoreline. This approach was used to quantify changes in both the ocean and sound shoreline between survey dates.

72. A straight-line shoreline length ℓ was used because the average shore-normal rate of change in shoreline position was desired. Generally, the ocean and sound shoreline orientation, α (Figure 27), did not vary at any site by more than a degree during the study period. This indicates that shore-line changes within 1-minute increments were mostly shore-normal; i.e., the coastline in the interval did not pivot a great deal. Therefore, the length ℓ between latitude or longitude boundaries 1 minute apart remained almost constant. The use of the straight-line distance ℓ rather than the actual shoreline distance was preferred on the sound side because (a) that shore was often very irregular and (b) one objective of the study was to compare ocean and sound shoreline changes. The sound shoreline change must, therefore, be viewed as the average rate of shore-normal movement based on changes in plan area and including nearshore islands. The straight-line shoreline length ℓ is thus a fairly constant, easily measured, and reasonable scaling factor to transform changes in area to shore-normal shoreline changes.

73. Areas, as shown in Figure 27, were digitized at NOS for each 1-minute increment. In the sound, islands immediately off the coast were included in the area computations because they had often been part of the coast at an earlier time; the islands were included only when they were clearly near the barrier island and when the sound beyond the island was open and wide.

74. The least-squares shoreline change rate S_ℓ is the slope of the best fit line to a plot of shoreline positions A_i/ℓ (Equation 1) versus time of each of the surveys in that 1-minute shoreline reach, or

$$S_\ell = \frac{\displaystyle\sum_{i=1}^{n} (t_i - \bar{t}) \frac{A_i - \bar{A}}{\ell}}{\displaystyle\sum_{n=1}^{n} (t_i - \bar{t})^2} \tag{2}$$

57

where

$$\bar{t} = \frac{1}{n} \sum_{i=1}^{n} t_i \tag{3}$$

and

$$\bar{A} = \frac{1}{n} \sum_{n=1}^{n} A_i \tag{4}$$

It is immaterial whether time is consistently taken as the date of the earlier or the later of the two surveys compared. The standard deviation SD of annual rates of shoreline change is

$$SD = \sqrt{\frac{1}{n-2} \sum_{i=2}^{n} (S_i - \bar{S})^2} \tag{5}$$

where

$$\bar{S} = \frac{1}{n-1} \sum_{i=2}^{n} S_i \tag{6}$$

Shoreline Change Rates

Listing of shoreline change rates

75. Shoreline changes, averaged (a) by varying shoreline distances (b) over the total survey period and parts of that period, are presented without interpretation in this section. Reasons for shoreline changes and relationships between shoreline changes and shelf bathymetry, inlets, capes, and shore orientation are discussed in the next section.

76. Tables 5-8 are listings of shoreline change rates for the period of approximately 1850 to 1980 for the following ocean shoreline reaches:

Table 5: Virginia, west of Cape Henry

58

Table 5

Ocean Shoreline Changes in Virginia West of Cape Henry*

Shoreline Longitude	1852-1859	1852-1980**	1852-1916	1859-1916	1916-1944	1944-1962	1962-1980
76°06'	1.3	-0.4		-1.2	0.1	-0.9	3.7
Lynhaven Inlet							
76°04'		1.0	0.8		1.8	0.6	0.8
76°03'		1.3	1.1		1.0	1.7	2.2
76°02'		1.7	0.5		3.3	1.6	3.7
76°01'	-6.8	-1.2		-1.5	-0.7	-0.6	1.1
76°00'		-1.1	-0.7		-3.2	0.5	0.4
75°59'		0.2	-0.5		1.1	0.2	2.0
Cape Henry							

* Shoreline change averaged between survey dates shown, in m/year; negative value indicates shoreline retreat.

** Least-squares estimate of shoreline change rate.

Table 6

Ocean Shoreline Changes in Virginia South of Cape Henry*

Shoreline Latitude	1852-1916	1852-1980**	1858-1925	1858-1980**	1859-1925	1859-1944	1859-1980**	1916-1944	1925-1942	1925-1944	1925-1980	1942-1962	1942-1980	1944-1962	1962-1980
Cape Henry															
36°55'	-0.5	0.2						1.1						0.2	2.0
36°54'						0.5	0.7							0.4	2.9
36°53'					0.2		0.1			0.3	0.2			-0.9	1.2
36°52'					-0.1		0.0			0.2	0.1			-0.5	0.7
Virginia Beach															
36°51'					-0.3		-0.2		-0.3		-0.5	-0.5			1.4
36°50'					0.0		0.0		-1.7		0.2	0.6			1.4
36°49'					-0.6		-0.5		0.5		-0.4	-0.9			-0.8
Rudee Inlet															
36°48'					-1.0		-0.7		-0.5		-0.1	0.1			
36°47'					-1.0		-0.7		-1.2		-0.2	0.3			
36°46'					-0.9		-0.9		-2.3		-0.8	-0.1			
36°45'					-0.8		-1.1		-3.7		-1.4	-0.3			
36°44'											-1.4				
36°43'															
36°42'					-2.3		-2.4				-2.5				
36°41'					-3.4		-3.0				-2.5				
36°40'					-2.3		-2.1				-1.9				
36°39'					-1.2		-1.5				-1.8				
36°38'					-1.3		-0.8				-0.1				
36°37'					-1.5		-0.9				-0.1				
36°36'															
36°35'			-0.4	-0.4							.3				
36°34'			-0.5	-0.2							.2				
36°33'			1.4	0.9							.2				
36°32'			1.3	1.0							.6				
36°31'			2.0	1.3							.5				
36°30'			1.6	1.1							.4				

* Shoreline change averaged between survey dates shown, in m/year; negative value indicates shoreline retreat.

** Least-squares estimate of shoreline change rate.

Table 7
Ocean Shoreline Changes in North Carolina North of Cape Hatteras*

							Survey Dates							
Shoreline Latitude	1849-1872	1849-1915	1849-1980**	1852-1946	1852-1980**	1872-1915	1915-1949	1915-1980	1946-1975	1949-1963	1949-1980	1963-1975	1963-1980	1975-1980
Southern Shores														
36°09'				0.0										
36°08'				0.1										
36°07'				-0.2										
36°06'		0.0	-0.5					-1.1						
36°05'		0.1	-0.6					-1.4						
Kitty Hawk Beach														
36°04'		0.7	-0.6					-1.9						
36°03'		0.7	-0.5					-1.7						
36°02'		1.5	-0.4					-2.3						
Croatan Shores														
36°01'								-1.7						
36°00'								-2.3						
35°59'		1.2	-0.7				-4.0	-2.3			-0.4			
35°58'		-0.2	-0.8				-1.7	-1.3			-0.9			
35°57'							-1.7	-1.4			-1.1			
Nags Head														
35°56'							-1.2	-1.3			-1.4			
35°55'		-1.1	-0.7				-0.2	-0.5			-0.9			
35°54'		-1.3	-0.7				0.1	0.3			0.5			
35°53'		-1.4	-1.2				-0.9	-1.0			-1.1			
35°52'		0.0	-0.7				-1.8	-1.3			-0.7			
35°51'		-0.8	-0.8				-0.8	-0.9		1.2			-2.8	
35°50'		-2.0	-1.5				0.0	-1.3		-3.3			-2.3	
35°49'		-3.6	-3.0				-0.8	-2.7		-6.3		-3.2		-3.9
Oregon Inlet														
35°45'										0.1		-4.3		-7.0
35°44'				-2.8	-2.9					-2.2			-4.9	
35°43'				-1.9	-1.8					-0.1			-3.1	
35°42'				-1.7	-1.3					1.0			-1.2	
35°41'				0.1	0.0					0.5			-1.7	
35°40'				0.0	-0.2					-1.4			0.5	
35°39'										-0.1			4.8	
35°38'							-3.2	-1.0		0.1			1.7	
35°37'		-3.7	-3.8				-4.5	-3.3		-5.3			0.7	
35°36'		-1.5	-3.2				-4.6	-4.5		-7.1			-2.2	
35°35'		0.3	-1.6				-1.9	-4.3		-2.0			-11.2	
Rodanthe														
35°34'		-0.4	-0.2				-0.1	-0.2		1.1			-1.4	
35°33'		0.4	0.3				-0.8	0.4		3.2			0.4	
35°32'		0.2	-0.3				-1.0	-0.4		-2.7			2.6	
35°31'		-0.6	0.0				-0.2	0.7		1.3			2.1	
35°30'		0.2	0.5				0.9	0.8		0.2			1.2	
35°29'		1.5	0.2				-1.5	-0.8		-2.5			1.9	
35°28'		1.4	0.1				-1.7	-0.8		-2.5			2.1	
35°27'		-0.3	-0.3				0.5	-0.1		-2.6			0.8	
35°26'		1.2	-0.5				-2.4	-2.0		-2.6			-0.7	
35°25'		1.5	-0.5				-4.1	-2.3		-1.3			0.5	
Little Kinnakeet														
35°24'		2.0	-0.3				-3.9	-2.6		-1.6			-0.8	
35°23'		5.0	1.2				-5.8	-2.5		0.3			1.8	
35°22'		4.0	0.2				-8.5	-3.9			1.2			
35°21'		0.8	-0.5					-1.9						
Avon														
35°20'		-0.1	-0.3				-1.3	-0.6			0.1			
35°19'		0.3	-1.0				-3.2	-2.0			-1.1			1.2
35°18'		-0.4	-2.0				-5.6	-3.7			-1.4			0.0
35°17'		-2.4	-3.3				-5.2	-4.2			-3.2			-2.2
35°16'	-2.5		-4.2			-5.3	-3.7	-3.5			-3.2			
35°15'	-10.2		-5.2			-6.3	-4.0	-3.2			-2.3			-1.9
35°14'														
Cape Hatteras														

* Shoreline change averaged between survey dates shown, in m/year; negative value indicates shoreline retreat.
** Least-squares estimate of shoreline change rate.

Table 8
Ocean Shoreline Changes West of Cape Hatteras*

Shoreline Longitude	Survey Dates						
	1860-1872	1860-1980**	1872-1917	1917-1946	1946-1963	1963-1975	1975-1980
Cape Hatteras							
75°32'	-1.2	7.6	11.5	10.1	-13.5	29.0	-8.1
75°33'	7.0	5.6	5.7	10.5	-5.4	2.7	22.9
75°34'	24.0	8.6	2.9	15.7	10.4	-6.0	21.7
75°35'	9.6	1.2	-1.8	5.9	-2.3	-1.2	2.3
75°36'	3.5	-0.3	-1.3	-0.8	1.6	-0.4	3.2

* Shoreline change averaged between survey dates shown; in m/year, negative value indicates shoreline retreat.
** Least-squares estimate of shoreline change rate.

Table 6: Virginia, south of Cape Henry

Table 7: North Carolina, north of Cape Hatteras

Table 8: North Carolina, west of Cape Hatteras

For the same period, Tables 9 and 10 list shoreline change rates for the following soundside shoreline reaches:

Table 9: Cape Henry to Cape Hatteras

Table 10: Pamlico Sound, west of Cape Hatteras

Ocean shoreline change rates

77. 1850 to 1980. The mean rate of change for the ocean shoreline over the approximately 130-year study period is shown in Figure 28. For those ocean reaches without rates shown, either no shoreline change values were available (i.e., the Corrolla to Duck, North Carolina, reach), or a major change in shore orientation (i.e., Capes Henry and Hatteras) or a break in the barrier island system (i.e., Oregon Inlet) precluded the determination of a usable ocean shoreline change rate. Between about 1850 and 1980 where data were available, approximately 28 percent of the ocean shore prograded, 68 percent retreated, and 4 percent did not change position.

78. Average shoreline change rates should be used with caution for planning and design purposes because large temporal and spatial variations in the rates have occurred in the past and can be anticipated to occur in the future. The standard deviation of shoreline position changes with time is a measure of these temporal variations. Large standard deviation values indicate a large variability in shoreline change rates between different surveys; smaller values indicate the shoreline change rate has been more nearly constant from one survey interval to the next. Figure 29 shows the standard deviation and the number of surveys used to calculate it for the east-facing ocean shore. Shoreline changes north of Oregon Inlet were relatively constant from 1852 to 1980 when compared to the changes south of Oregon Inlet to Cape Hatteras. Greater variations in shoreline position are the norm for the latter 60-km-long reach.

79. Partial study period. Dates of survey allow a separation of the data set into two nearly equal time intervals. It is useful to compare ocean shoreline changes for those two periods for several reasons. During the period from about 1850 to 1915-1925, the shoreline underwent mostly natural changes, except for the dune vegetation loss caused by grazing animals. During the period from 1915-1925 to 1980, human intervention in the form of

62

Table 9

Sound Shoreline Changes, Cape Henry to Cape Hatteras*

Shoreline Latitude	Survey Dates																	
	1849- 1915	1849- 1980**	1852- 1917	1852- 1946	1852- 1980**	1857- 1980**	1858- 1980**	1859- 1980	1915- 1949	1917- 1946	1917- 1949	1946- 1963	1946- 1980	1949- 1963	1949- 1980	1963- 1975	1963- 1980	1975- 1980
36°42'								0.3										
36°41'								-1.2										
36°40'								-6.4										
36°39'								-10.8										
36°38'								-4.1										
36°37'								-7.5										
36°36'†								-1.5										
36°34'								-0.1										
36°33'							-0.4											
36°32'							1.0											
36°31'							-4.4											
36°30'							-0.4											
36°28'						0.9												
36°27'						2.3												
36°26'						-1.0												
36°25'						0.1												
36°24'						1.3												
36°21'					/ -0.2													
36°20'					0.0													
36°19'					-1.6													
36°18'					1.0													
36°17'					2.2													
36°16'					1.3													
36°15'					-1.3													
36°14'					-0.9													
36°13'					-0.1													
36°08'					-0.3													
36°05'		0.1																
36°04'		-0.3																
36°03'		-1.2																
36°02'		3.1																
35°58'	-0.7	-0.8						-1.4							-0.1			
35°57'	-0.2	0.2						0.9							0.2			
35°56'								-1.2							-1.0			
35°55'	2.3	1.4						0.8							-0.9			
35°54'	-0.4	-0.3						-0.4							0.1			
35°53'	-2.8	-1.5						0.1							-0.6			
35°52'	2.6	1.3						-0.1							-0.2			
35°51'	13.1	6.8						0.8						-0.4			-0.1	
35°50'	3.9	2.4						1.6						-0.7			0.2	
35°49'	0.2	-0.1						-0.3						-1.2			0.3	
Oregon Inlet														-0.5				
35°45'															-1.4	-5.8		12.2
35°44'			1.3	1.1								0.7					0.0	
35°43'			1.5	1.0								-0.6					-1.3	
35°42'			5.3	4.2								0.1					-0.4	
35°41'			11.1	8.7								-0.7					-1.1	
35°40'			4.4	3.0								-3.5					-1.5	
35°39'												-2.2	-2.0				-1.8	
35°38'								21.5	1.0								-1.5	
35°37'			4.0	2.4				1.2		-0.3							-0.7	
35°36'			1.4	0.3				-1.3		0.2							-0.6	
35°35'			-0.8	-0.4				0.1		0.2							-0.6	
35°34'			0.1	-0.1				-0.4		-0.4							-0.1	
35°33'			-0.1	-0.4				-1.0		-0.3							-0.1	
35°32'			0.4	-0.4				-1.9		-0.5							0.2	
35°31'			0.6	-0.4				-2.3		-0.7							0.3	
35°29'			-0.8	-0.4				0.3		-0.1							-0.3	
35°28'			-0.7	⌐0.6				-0.4		-0.5							-0.2	
35°27'			-0.9	-0.8				-0.9		-0.5							-0.5	
35°26'			-1.2	-0.9				-0.7		-0.8							-0.4	
35°25'			-1.0	-0.8				-0.6		-0.9							0.0	
35°24'			-1.0	-0.6				0.0		-0.3							0.2	
35°23'			-1.4	-1.0				-0.6		-0.8							0.0	
35°22'			-0.9	-0.8				-1.9							0.9			
35°21'			-0.8	-1.3				-2.0							-1.7			
35°20'			-1.2	-1.1				-1.2							-0.9			
35°19'			-1.3	-1.0				-0.7							-0.4			
35°17'			-0.2	-0.1				-0.5							0.8			

* Shoreline change averaged between survey dates shown, in m/year; negative value indicates shoreline retreat.
** Least-squares estimate of shoreline change rate.
† Some latitudes not included because data were unavailable.

63

Table 10

Pamlico Sound Shoreline Changes West of Cape Hatteras*

Shoreline Longitude	Survey Dates				
	1860–1872	1860–1980**	1872–1917	1917–1946	1946–1980
75°32'				-0.5	-0.3
75°33'	-0.9	-1.1	-1.0	-1.4	-0.8
75°34'	4.1	-2.5	-4.1	-3.1	-0.6
75°35'	-0.5	-1.5	-1.7	-1.7	-1.2
75°36'	2.1	-1.5	-4.0	1.6	-1.4

* Shoreline change averaged between survey dates shown, in m/year; negative value indicates shoreline retreat.
** Least-squares estimate of shoreline change rate.

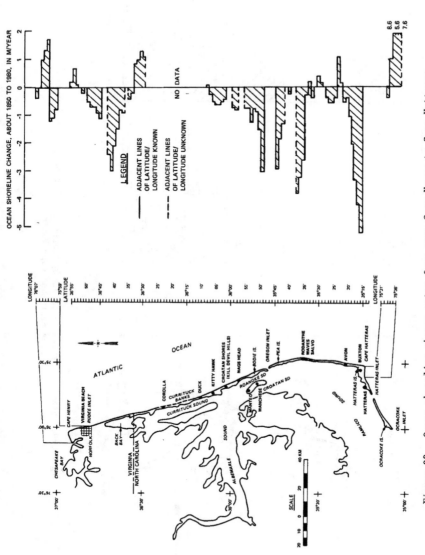

Figure 28. Ocean shoreline change rates from near Cape Henry to Cape Hatteras, from about 1850 to 1980 (least-squares shoreline change rate is the slope of the best fit line to shoreline position versus time of each survey in each 1-minute reach)

65

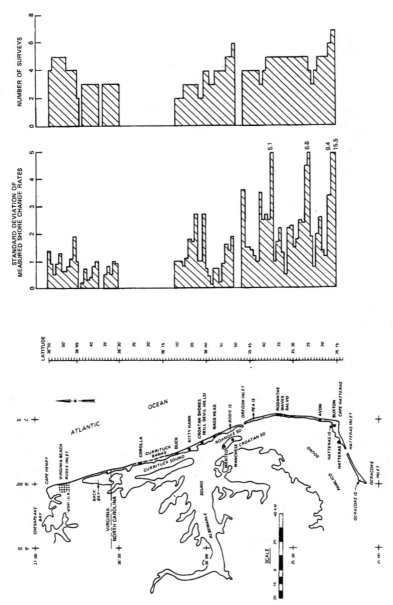

Figure 29. Standard deviation of ocean shoreline position changes between Cape Henry and Cape Hatteras

66

artificial dune-building occurred along much of the ocean shore; it was also in this latter period that recreation became the dominant industry in the area. Figure 30 shows average ocean shoreline change rates for the two periods in the 36-km-long coastal reach south of Cape Henry; Figure 31 illustrates the same parameters for the reach between Duck and Cape Hatteras, North Carolina. The study area was divided into two areas because of the long intervening reach for which shoreline position data were unavailable. Shoreline change rates at Virginia Beach, Virginia, for the most recent 55 years of survey data are illustrated in Figure 32; they are included separately because the influence of man in recent times at Virginia Beach has increased significantly.

80. Extreme shoreline position excursion is another measure of shoreline change variability (Figure 33). Taken over the 130-year period of record this provides the maximum landward and seaward position to which the shoreline moved, based on infrequently-taken survey data; the actual extreme shoreline

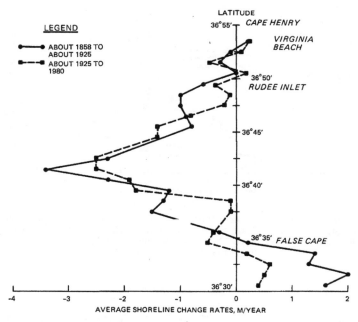

Figure 30. Average ocean shoreline change rates for the 36-km-long reach south of Cape Henry in the periods 1859-1925 and 1925-1980

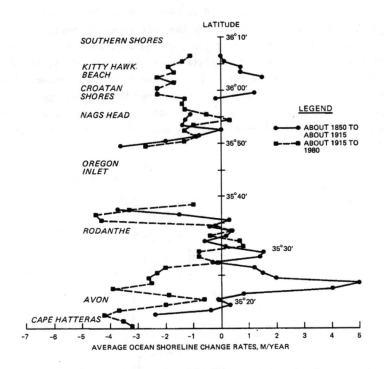

Figure 31. Average ocean shoreline change rates for two survey periods (about 1850 to about 1915 and about 1915 to 1980) in the reach between Duck, N. C. (latitude 36°06'), and Cape Hatteras (latitude 35°15')

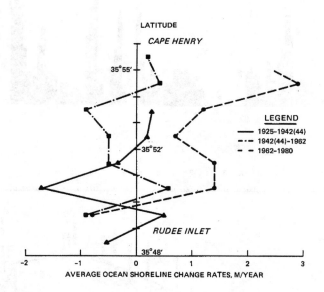

Figure 32. Average ocean shoreline change rates at Virginia Beach for four successive surveys between 1925 and 1980: (a) 1925-1942(44), (b) 1942(44)-1962, and (c) 1962-1980

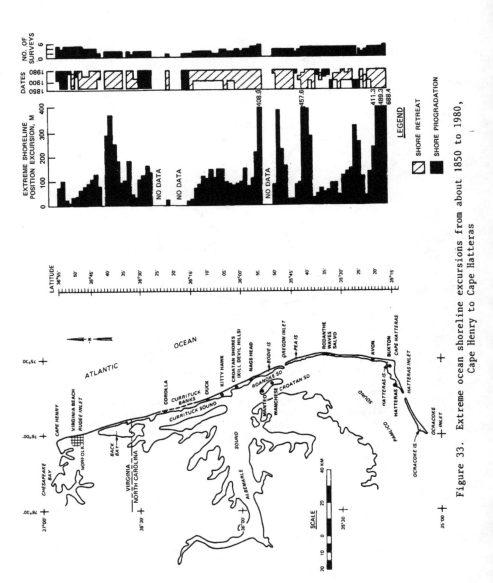

Figure 33. Extreme ocean shoreline excursions from about 1850 to 1980,
Cape Henry to Cape Hatteras

70

excursion probably was larger than appears because only a few shoreline positions were measured. In over one-half the surveyed shore reach, the extreme shoreline position change occurred between the first and last surveys, indicating a relatively continuous shore retreat or advance. Because of this trend, patterns shown in Figure 33 are similar to mean shoreline change rate patterns shown in Figure 28. Areas of retreat are more numerous than areas of advance.

Sound shoreline change rates

81. <u>1850 to 1980.</u> Shoreline changes on the sound side of the barrier islands exhibit few consistencies in an alongshore direction (Figure 34). The largest retreat rates occurred in the Back Bay area, a constricted freshwater region reported to be free of inlets in historic time (Figure 9). Accretionary trends adjacent to and south of Oregon Inlet appear to be inlet-associated. The consistent 0.5- to 1.5-m/year retreat of the sound shoreline south of Salvo, North Carolina, also occurs in an area where inlets persistent in historic times have not been reported. Figure 35 illustrates the standard deviation of shoreline position change in the sounds through time.

82. <u>Partial study period.</u> Figure 36 shows changes in the sound shoreline for the same periods illustrated for the ocean shoreline in Figure 31. The 1852-1980-averaged sound shoreline change rate was -0.1 m/year, a retreat which is 13 percent of the average retreat rate (-0.8 m/year) of the ocean shore.

83. In the sound adjacent to Oregon Inlet the standard deviation of shoreline change (Figure 35) is very large, suggesting fluctuations that are likely inlet-related. South of Rodanthe, the shoreline retreated in a relatively continuous manner. Infrequent and severe storms probably caused the changes in this area. However, the storm effects, which are usually localized in time and location, were probably lost in spatial averaging, especially considering the large survey interval of this study.

Oregon Inlet

84. Oregon Inlet was opened in 1846 by a severe coastal storm. Initially, large quantities of water moved through New Inlet (Figure 9) into Pamlico Sound. Precipitation and runoff further increased the volume of water in the sound; when the wind direction changed to the west, some of this ponded water was carried seaward north of the site of present-day Oregon Inlet

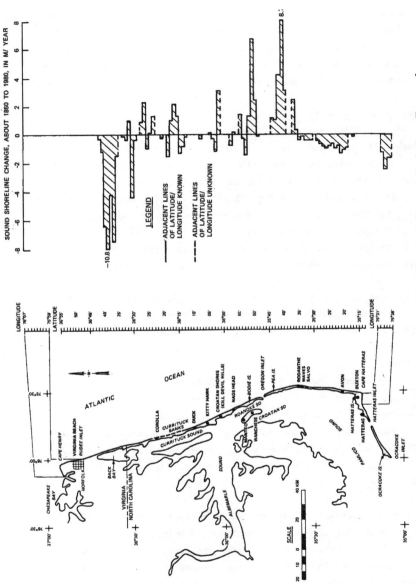

Figure 34. Sound shoreline change rates for the reach between Back Bay, Va., and 12 km west of Cape Hatteras, from about 1850 to 1980 (least-squares shoreline change rate is the slope of the best fit line to shoreline position versus time of each survey in each 1-minute reach)

72

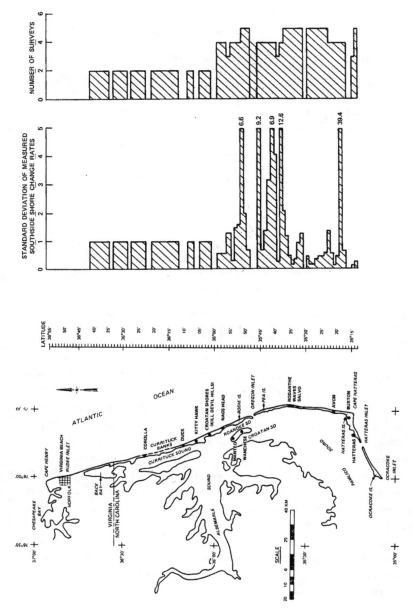

Figure 35. Standard deviation of sound shoreline position changes between Cape Henry and Cape Hatteras, from about 1850 to 1980

73

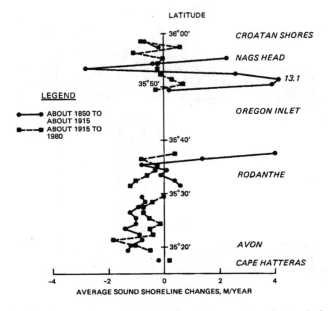

Figure 36. Average sound shoreline changes, for the periods from about 1850 to about 1915 and from about 1915 to 1980, between Nags Head and Cape Hatteras, N. C.

(Cumming 1966). With time, a channel was cut and deepened. Tidal currents through the inlet throat have since kept it from filling with littoral sediments carried in a shore-parallel direction.

85. The inlet has not remained fixed in its original position, nor has its shape nor the shape of the adjacent islands remained constant. Figure 37 shows the changes that have occurred since the inlet opened. (The dashed line is the 1849 shoreline included for comparison purposes; note that a short reach of shoreline south of Oregon Inlet was not surveyed during the 1915-1917 period.) Between about 1849 and 1980 the average inlet migration rate (i.e., the shore-parallel (north-south) movement of the midpoint of the narrowest part of the inlet throat) to the south for Oregon Inlet was approximately 29 m/year. As shown in Figure 38 this rate varied greatly from one survey interval to the next. The most rapid movement of the midpoint, 87.5 m/year south between 1963 and 1975, occurred just after a severe storm on 6 and 7 March when the inlet widened and migrated north.

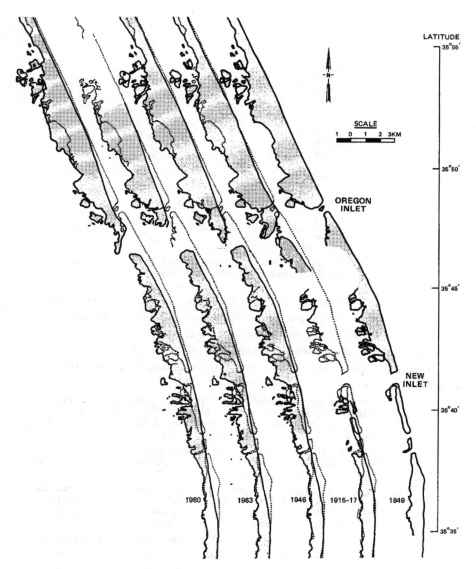

Figure 37. Changes in ocean and sound shorelines adjacent to
Oregon Inlet, N. C., for five surveys between 1852 and 1980

75

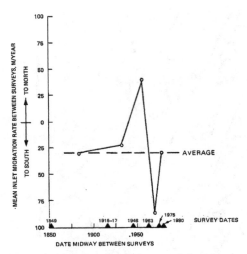

Figure 38. Migration rates of Oregon Inlet throat for five
survey intervals between 1849 and 1980

86. Figure 39 shows the relative locations and orientations of the
narrowest section of the Oregon Inlet throat measured during six surveys
between 1849 and 1980. Also shown are changes in narrowest inlet throat
width, the relative location of the center of the inlet throat, and the direction of inlet throat migration. This figure emphasizes three interesting
features:

> <u>a</u>. The width of the inlet throat in 1963, about 2.5 km, was over
> twice as large as the width average which is about 1.2 km. The
> throat was expanded during the storm of March 1962, mostly at
> the expense of the island to the north.

> <u>b</u>. With the exception of the 1962 storm period (1963 survey) when
> the center of the inlet throat moved north, migration was in a
> generally south direction.

> <u>c</u>. Except for immediately after the 1962 storm, the orientation
> of the channel at the narrowest section was approximately
> north-south; i.e., a line connecting the two sides of the
> inlet at its narrowest section was oriented east-west.

87. The change in land area adjacent to Oregon Inlet very likely reflects the inlet influence on the nearby shorelines. This land area (Figure 40) above mean high water has been declining since the inlet opened;
with the exception of the 1963-1980 interval, the loss has averaged about

76

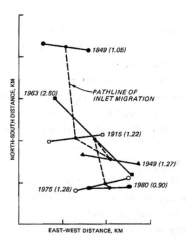

Figure 39. Relative locations and orientations of the narrowest section of Oregon Inlet throat, 1849-1980 (dates given are dates of survey; numbers in parentheses are inlet widths in kilometers)

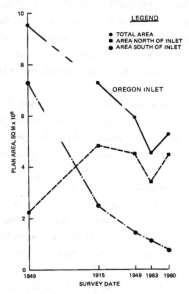

Figure 40. Plan view changes in land area in the vicinity of Oregon Inlet, N. C., 1849-1980

77

-36,000 sq m/year. The land region included in the analysis extended 8 km north and 8 km south of the inlet as it existed in 1849 (Figure 37). The net loss for the total system, which may include some effects of New Inlet, resulted from shoreline retreat adjacent to the inlet; this probably represents a transfer of sand from the ocean shoreline region to the sound by inlet currents. Note that the decrease in plan area was continuous south of the inlet but the variable plan area changed north of the inlet. The increase to the north occurred as the result of a spit which built south as the inlet migrated south.

Cape Hatteras and Cape Henry

88. The land protrusion of Cape Hatteras (Figure 2) has changed significantly in plan area and shape since the first survey in 1852 (Figure 41). Figure 42 shows that a decrease in land area occurred as the east-facing coast retreated (eroded) to the west and the south-facing coast prograded (accreted). Figure 28 shows evidence of the ocean shoreline retreat at the Cape Hatteras lighthouse. In 1870, when the lighthouse was built, the shoreline was 600 m east of its 1980 position. Retreat has been continuous except for a brief period in the 1940's where a slight progradation occurred. Figure 43 shows that Cape Point moved about 0.5 km in a net southwesterly direction between 1852 and 1980. The figure also shows that Cape Point fluctuated greatly in position during that period and that the present position is likely a temporary site.

89. Cape Henry, during the same period, changed in a different way (Figure 44). The east-facing shore moved east (prograded), while the north-facing Chesapeake Bay shore moved south (eroded). The changes at Cape Henry were nearly an order of magnitude smaller than the changes at Cape Hatteras.

Variations in shoreline change rates with time

90. Shoreline change rates varied greatly with time. The extent of this variation is illustrated in Tables 11 and 12. The periods shown on the tables, 1852-1917, 1917-1949, and 1949-1980, are expedients based on available survey data. The shoreline change data, when averaged by reach (Figures 45 and 46) suggest these trends:

> a. Shoreline retreat on the east-facing ocean coast was at a maximum during the 1917-1949 period (Figure 45). Greatest shore stability occurred between 1852 and 1917.

78

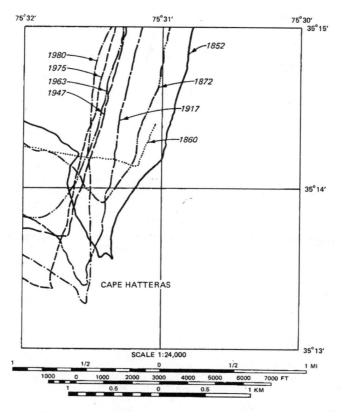

Figure 41. Changes in the mean high-water shoreline at Cape Point, Cape Hatteras, between 1852 and 1980 (dates indicate the year of the shoreline survey)

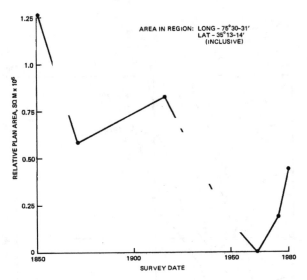

Figure 42.   Relative plan area of the subaerial projection for
Cape Hatteras, 1852-1980

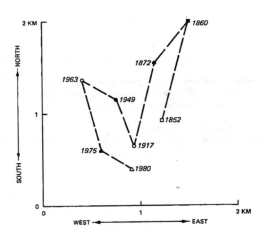

Figure 43.   Location of Cape Hatteras point between 1852 and 1980

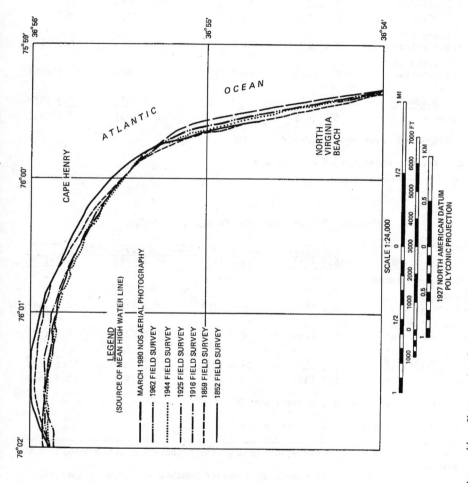

Figure 44. Changes in the mean high water shoreline at Cape Henry between 1852 and 1980

81

Table 11

Summary of Mean Shoreline Changes, Oceanside

| Coastal Reach | Mean Shoreline Change for These Survey Periods, m/year* | | | |
	1852-1917	1917-1949	1949-1980	1852-1980
West of Cape Henry	0.2 (5)**	0.5 (7)	1.2 (7)	0.2 (7)
Cape Henry to Oregon Inlet†	-0.0 (20)	-1.2 (18)	-0.3 (21)	-0.6 (40)
Oregon Inlet to Cape Hatteras††	0.4 (21)	-2.9 (23)	-1.3 (30)	-1.1 (28)
West of Cape Hatteras	4.5 (5)	8.3 (5)	2.0 (5)	4.5 (5)
Average	0.6 (51)	-0.8 (53)	-0.4 (63)	-0.4 (80)

Mean N-S ocean coast = -0.8 m/year; n = 75

Mean E-W ocean coast = +2.0 m/year; n = 12

Table 12

Summary of Mean Shoreline Changes, Soundside

| Coastal Reach | Mean Shoreline Change for These Survey Periods, m/year* | | | |
	1852-1917	1917-1949	1949-1980	1852-1980
West of Cape Henry	--	--	--	--
Cape Henry to Oregon Inlet	2.0 (9)**	0.1 (10)	-0.2 (10)	-0.5 (40)
Oregon Inlet to Cape Hatteras†	0.7 (24)	0.3 (20)	-0.2 (27)	0.3 (19)
West of Cape Hatteras‡	-1.9 (4)	-1.0 (5)	-0.9 (5)	-1.7 (4)
Average	0.7 (37)	0.1 (35)	-0.3 (42)	-0.3 (63)

Mean N-S sound shore = -0.1 m/year; n = 70

Mean E-W sound shore = -1.2 m/year; n = 5

* Positive = shoreline moves seaward; negative = shoreline moves toward mainland.
** Number in parentheses is number of 1-minute reaches included in analysis; number varies for different time periods because surveys are not continuous for entire coast.
† Data coverage = approx. 60 percent of total shoreline.
†† Data coverage = approx. 90 percent of total shoreline.
‡ Data coverage (1852-1980) = approx. 80 percent of total shoreline.

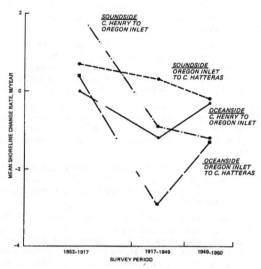

Figure 45. Shoreline change rates, averaged by survey period, for east-facing ocean shorelines and west-facing sound shorelines

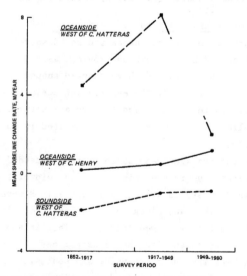

Figure 46. Shoreline change rates, averaged by survey period, for west-facing ocean and sound shorelines

83

<u>b</u>. The temporal trend, but not the magnitude, of east-facing ocean
coast changes was similar north and south of Oregon Inlet
(Figure 45).

<u>c</u>. North- and south-facing ocean coasts (Figure 46) were accre-
tional in the 5- to 10-km study reaches west of the capes for
all survey periods. The trends of change were not similar;
however, the small number of reaches sampled in each area
(Table 11) may preclude a realistic comparison.

·<u>d</u>. The west-facing shoreline trend in the sounds was one of continuous change from progradation (movement into the sounds)
to retreat (movement toward the ocean) between 1852 and 1980
(Figure 45). The trends were similar north and south of Oregon
Inlet.

<u>e</u>. Between 1852 and 1980, the north-facing shoreline west of Cape
Hatteras decreased its net retreat (Figure 46). This trend was
the opposite of that measured for the west-facing sound shoreline (Figure 45).

<u>f</u>. Ocean and sound shoreline changes generally did not follow
similar trends through time. While the east-facing ocean
shoreline retreated at a maximum rate between 1917 and 1949,
the west-facing sound shoreline (i.e., the shoreline on the
other side of the barrier island) reached a maximum retreat
rate in the 1949-1980 period. Only the north-facing ocean
shoreline at Cape Henry and the north-facing sound shoreline
at Cape Hatteras (Figure 46) showed similar behavioral trends
through time.

Changes in island width and position

91. Where data covering both ocean and sound shorelines are available,
an analysis of island width and position provides useful information on the
particular ways in which the islands have changed shape. When averaged for
the period of about 1850 to 1980, the east-facing ocean shore retreated an
average 0.8 m/year. In the same period the average retreat rate of the west-
facing sound shoreline was 0.1 m/year. This resulted in an average island
narrowing of 0.9 m/year.

92. Because the average ocean shore retreat exceeded the average rate
of sound shore retreat, the island axis (i.e., the midpoint between.shore-
lines) moved landward (west) an average 0.35 m/year. However, as Table 13
shows, in most time periods and along most reaches, the island axis moved seaward at more locations than it moved landward. This island axis movement,
though, should not be confused with the classical definition of barrier island
migration which assumes that both oceanside and soundside shorelines move
toward the continental land mass. Island migration occurs when the ocean
shoreline erodes and, concurrently, the sound shoreline progrades as sand is

Table 13

Combined Ocean- and Soundside Shoreline Changes

| Survey Period 1852-1980 | Number of 1-minute Latitude/Longitude Shoreline Increments Which Moved | | | | |
	North of Oregon Inlet	South of Oregon Inlet	West of Cape Hatteras	Total	No Change
Island* widens	8	4	2	14	
Island narrows	16	15	2	33	
Island axis moves toward sound	12	7	0	19 ⎫	
Island axis moves toward ocean	10	12	4	26 ⎭	2
Survey Period 1852-1980					
Island widens	4	9	2	15 ⎫	
Island narrows	4	7	2	13 ⎭	3
Island axis moves toward sound	6	6	0	12	
Island axis moves toward ocean	2	13	4	19	
Survey Period 1946-1980					
Island widens	1	7	2	10 ⎫	
Island narrows	9	18	3	30 ⎭	
Island axis moves toward sound	9	12	0	21 ⎫	
Island axis moves toward ocean	0	12	4	16 ⎭	4

* Island as shown here also includes the peninsula or spit north of Oregon Inlet.

transported (a) across the island by overwash or wind or (b) through inlet openings directly to the sound shoreline.

93. A general indication of island behavior is shown in Table 13 for three regions (i.e., north and south of Oregon Inlet and west of Cape Hatteras) in the study area. Note that island narrowing, by portion of the coast, is the most common change, while island widening is the least common behavior. Slightly more segments of the island system moved seaward than landward. Table 13 references direction of movement by 1-minute latitude or longitude increments. The following four changes are noteworthy:

a. For the measured segments, island narrowing greatly exceeded island widening when averaged for the 130-year study period. However, during the 1852-1917 period, island widening and narrowing were almost equal. Between 1946 and 1980, three times as much of the measured island system narrowed as widened.

b. Island-narrowing-to-widening ratios were generally similar north and south of Oregon Inlet. This suggests that the conditions which led to the island width changes, while they varied through time, were consistent throughout the study area.

c. Over the study period, the island axis moved seaward at slightly more places than it moved landward. This situation, however, varied by survey period. Between 1852 and 1917 seaward movement prevailed, while between 1946 and 1980 landward movement of the axis prevailed.

d. For the measured portions of the study area, trends in island narrowing or widening did not indicate particular movements of the island axis.

94. Figures 47-50 show the rates of island width change and the rates of change in position of the island axis for different time periods. Figures 48 and 50 are limited to the section between Kitty Hawk and Cape Hatteras, North Carolina, because that is the only area in which data were available for both the periods 1852-1917 and 1917-1980. These figures illustrate the following alongshore changes in island width and position through time:

a. The largest island width changes occurred near Back Bay, Virginia (Figure 47). This is an area of large ocean (Figure 28) and sound (Figure 34) shoreline retreats.

b. Increases in island width between Kitty Hawk and Oregon Inlet (Figure 47) came as a result of progradation of the sound shoreline (Figure 34) during a period of ocean shore retreat (Figure 28) (the area near Croatan Shores was not influenced by an inlet during the study period). The south-facing ocean coast west of Cape Hatteras was also an area of island width increase; however, here the increase occurred because of ocean shore progradation (Figure 28).

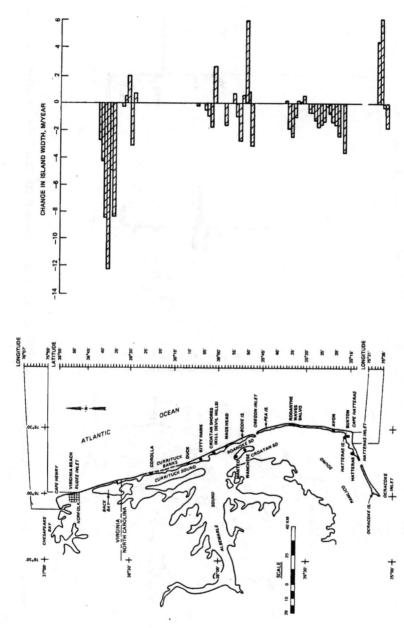

Figure 47. Island width changes between about 1850 and 1980, from Cape Henry To west of Cape Hatteras

87

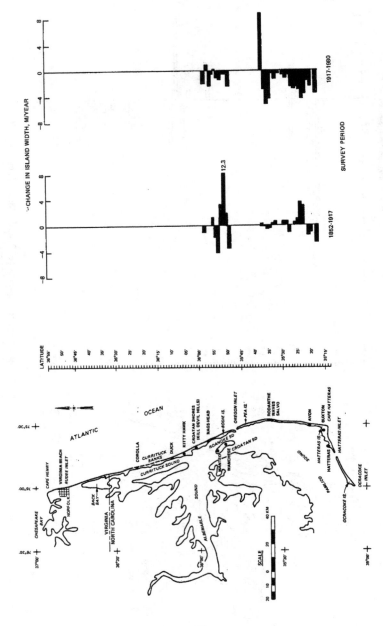

Figure 48. Island width change rates, 1852-1917 and 1917-1980, between Kitty Hawk and Cape Hatteras, N. C.

88

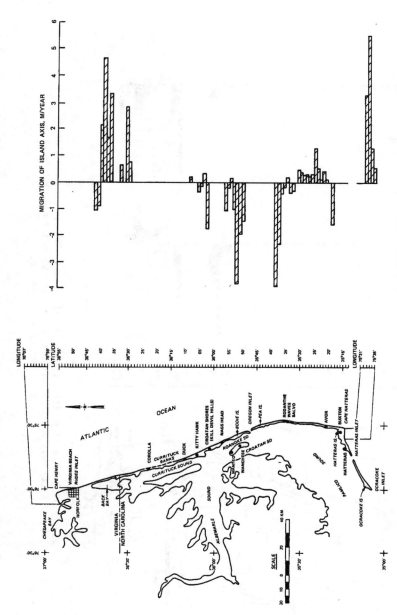

Figure 49. Rates of position change of island axis between about 1850 and 1980, Cape Henry to west of Cape Hatteras (island axis = midpoint between ocean and sound shorelines)

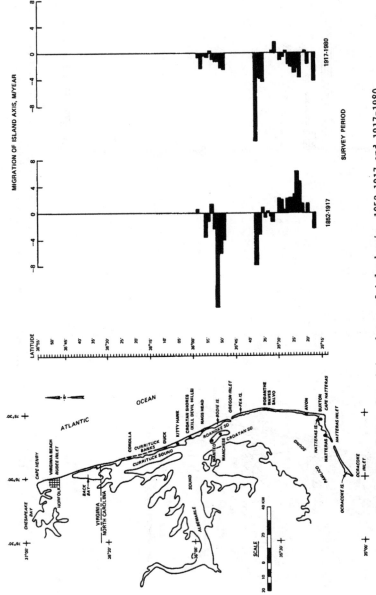

Figure 50. Rates of position change of island axis, 1852-1917 and 1917-1980, between Kitty Hawk and Cape Hatteras, N. C. (island axis = midpoint between ocean and sound shoreline)

<u>c</u>. Island narrowing south of Oregon Inlet (Figure 47) generally occurred as a result of combined ocean and sound shore retreat (Figures 28 and 34).

<u>d</u>. Island width changes varied greatly in time in both magnitude and direction (Figure 48). The period 1852-1917 was one of slightly greater island widening; between 1917 and 1980, island narrowing predominated.

<u>e</u>. Island axis migration rates (Figure 49) may be positive (seaward-moving) in Back Bay but, because island narrowing along both shorelines predominated here (Figure 47), to consider this axis migration as island migration is misleading. It is best thought of as island narrowing, with retreat of the sound shoreline greater than the ocean shoreline.

<u>f</u>. Changes directly north and south or Oregon Inlet are the result of inlet processes: the ocean shoreline has retreated and the sound shoreline has prograded (Figures 28 and 34). Processes associated with Oregon Inlet and New Inlet (Figure 9) are responsible.

<u>g</u>. Island migration was similar in direction for the 1852-1917 and 1917-1980 survey periods, with one exception (Figure 50): the island axis in the region centered on Avon, North Carolina, moved seaward in the former period and toward the mainland in the latter period. This change in direction is primarily the result of a shifting ocean shoreline (Figure 28).

91

PART V: PREDICTION OF FUTURE SHORELINE CHANGES

95. The NOS shoreline change maps show what happened between Cape Henry and Cape Hatteras from 1850 to 1980. An analysis of the maps quantifies the changes both spatially and temporally. Data regarding historical shoreline changes can provide useful information with which to predict future changes. When the causes of change are imperfectly known, however, it is difficult to predict future changes by extrapolating past trends because the future may not mimic the past. It is not apparent from the results of this study that the magnitude of future changes in shoreline behavior can be forecast. However, future changes at specific sites can probably be estimated for any given time period relative to the average changes which have occurred in the rest of the study area. This section treats these aspects of shoreline change prediction separately.

Temporal Predictions

96. Great variability was found in change rates within the 1850-1980 survey period. It is not unreasonable to assume future changes will be different from the 1850-1980 average. The survey record of shoreline changes in the study area is relatively short, intermittant, and nonuniform in frequency; it also lacks noticeable trends through time (Tables 11 and 12, and Figures 45 and 46). Consequently, there is limited shoreline change data available with which to extrapolate shoreline changes into the future. In addition, because of the multiplicity of processes involved, it is impossible to evaluate the relative importance of man's impact relative to changes in the natural processes that caused the shore to accrete or erode.

Spatial Predictions

97. Many changes in shoreline position are likely related to local conditions. Because wave, wind, and current data are unavailable over the 130-year survey period and throughout the study reach, a direct causal relationship cannot be established to predict those changes. However, most of the alongshore variations in shoreline change appear to be influenced by the proximity of the shoreline to inlets, capes, and nearby shore-connected ridges

92

(Figure 51). The relationship between shoreline change and these features appears reasonable and informative, but the relationship does not consider the actual processes causing the changes. Extrapolation of future shoreline changes using both past shoreline change data and the relationships between those changes and local features improves the forecasts, but even these predictions must be treated with caution. Clearly, an effort to establish the causes of the shoreline changes related to local features is warrented.

Barrier island migration and narrowing

98. Barrier islands along the mid-Atlantic coast very likely formed on the Continental Shelf considerably east of their present positions during a period when sea level was much lower than it is today (Swift et al. 1972). As sea level rose, the islands are thought to have migrated toward the continental land mass--or west in the study area. For this migration to have occurred, the ocean side of the islands must have retreated and the sound side must have prograded. During migration the islands likely had alternating periods of net island narrowing and widening superimposed on the longer term landward migration. Conditions favoring island migration are those that move sand from the ocean side of the islands to the sound side. In the study area, this would mean one or more of the following conditions:

a. Overwash transport. The optimum conditions are a narrow island (probably less than 1 km in width, and maybe quite a bit less); a low island where dunes are absent, or low and discontinuous; minimum vegetation, especially those shrubs and trees that would hinder overwash; and storm surges of long duration in which the water level exceeds the island elevation.

b. Aeolian transport. The optimum condition is a strong onshore wind that exceeds 25 km/hour (that necessary for sand transport) for long periods of time; a wide, dry beach area that serves as a source for wind-carried sand; and an absence of vegetation so that the windblown sand can be carried to the sound side of the island. (A low, narrow island would probably allow a more speedy trip for a sand grain from ocean to sound but is not necessary for effective aeolian sand transport.)

c. Inlet-related transport. Most important to island migration is the presence of many large and relatively permanent inlets which intercept sand moving in the littoral zone and move it in a net westward direction. An inlet is capable of removing a large portion of the sand moving in an alongshore direction and transferring it to shoals in the sound or to the sound shoreline adjacent to the inlet. As the number, size, and persistence of the inlets increase, the amount of sand moved in a landward direction increases and the probability of island migration increases.

93

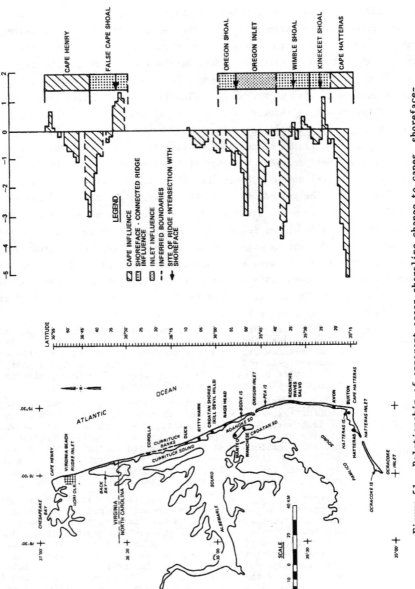

Figure 51. Relationship of apparent ocean shoreline change to capes, shoreface-connected ridges (shoals), and Oregon Inlet

94

99. The data presented herein have shown that although islands in the study area narrowed over the 130-year study period (Figures 47 and 48), they did not migrate in the classic sense toward the mainland because both the ocean and sound shorelines retreated toward the island. The reasons island migration has ceased are not clear. Quite likely, overwash has not been an important mechanism in sound shoreline progradation for the last several hundred years. Today, the islands are probably too wide in most places for overwash penetration across the entire island (Leatherman and Fisher 1976). In addition, prior to about 1800 the islands were well vegetated with trees and shrubs (Hennigar 1979) which would have either inhibited overwash or been destroyed had frequent or severe overwash conditions existed. In the nineteenth and early twentieth centuries, aeolian transport may have been of some local significance because poor land practices had left the island barren (Hennigar 1979). However, at other times wind-transported sand probably did not account for much sound shoreline progradation. If island migration occurred in the study area between 1585 and 1850, it was probably the result of inlet processes. Figure 9 shows that the number and permanency of inlets have decreased in the study area from 1585 to the present time; if migration is not occurring today, it is probably because the impact of inlets is too small. Only Oregon Inlet now acts as a sediment trap in the study area; significantly, the barriers adjacent to it are migrating in a westerly direction.

100. The reasons for island narrowing are also not clear; nor is it clear when the narrowing cycle began or when it will end. Sand losses from the front and back of the islands in the recent past may have been partially caused by a rise of sea level relative to land--a vertical rise of probably 4 mm/year in the study area since 1930 (Hicks 1981) (on a static shore slope of 1:40, for example, this would translate to an apparent shore retreat of 0.1 m/year). Quite likely a relative sea level rise would also have caused dynamic changes in the beach that would have increased the shore retreat rate; this effect cannot be quantified at present. Long-term changes in wave and wind conditions also could have forced the ocean and sound shores to retreat or accrete, especially if the frequency and duration of storms had changed substantially. An added factor, frequently not considered, is that unconsolidated marine coasts may retreat under "normal" conditions. Whether the rate of relative sea level rise will increase or decline and whether wind and wave conditions will produce more or less erosion in the future are unknown.

101. Island narrowing must have begun before 1850. It will, of course, end when the islands disappear or when one or both shorelines begin to prograde. At the present average rate of island narrowing (0.9 m/year), it will take almost 1700 years for a 1500-m-wide island to narrow to nothing. Before that happens, though, overwash, if allowed, will likely begin to transport sand to the sound shoreline and island migration will commence. A reasonable forecast based on past behavior is that narrowing will probably continue in the foreseeable future.

Alongshore sediment transport reversal

102. Waves approaching shore at acute angles and winds with a shore-parallel component create alongshore currents. Sediment mobilized by wave activity is moved by these currents. Over the period of a year the amount of sand moved one way is rarely balanced by that which is moved the other way; the difference is the net volume of littoral sand which moved preferentially in one direction. This net volume and the direction it is moved may change from year to year and over longer time periods as the wave and wind climate changes.

103. Study results tell us little about the net volume moved; however, they provide some indication of the direction of net sediment transport. Other studies have suggested that, on the long-term average, 2×10^5 cu m/year of sediment moves north at Rudee Inlet* and 5×10^5 cu m/year moves south at Oregon Inlet (U. S. Army Engineer District, Wilmington 1980). This net transport indicates that a change in the net alongshore sediment transport direction-- i.e., a transport reversal--occurs somewhere between the two inlets. Evidence from this study suggests that the reversal occurs near latitude 36°41' (Figure 28 shows a very uniform decrease in the shoreline retreat rate north and south of that site (the north end of Back Bay)) to create a divergent longshore sediment transport nodal zone; i.e., a place where sand moves alongshore to both the north and the south away from the site. Losses north and south of latitude 36°41' are nearly equal and decrease progressively with distance. Shoreline retreat rates are expected to decrease if a divergent nodal zone exists because sediment moving away from the node will reach adjacent beaches and thereby reduce the loss rates there.

104. The large shoal complex east of the Chesapeake Bay entrance

---

* Personal Communication, James Melchor, 1981, Oceanographer, U. S. Army Engineer District, Norfolk, Va.

influences the refraction path of waves approaching the coast from north to east. The effect of the topographic high is to bend waves approaching from north of shore-normal to approach from the south. This mechanism tends to create a northward-directed current, which supports the inference that an alongshore sediment transport nodal zone exists near latitude 36°41'.

105. At Rudee Inlet, the net alongshore transport rate of 2×10^5 cu m/ year based on recent dredging records, is 60 percent of an estimated 3.4×10^5 cu m/year cummulative volume loss north of the nodal zone. The latter value is based on long-term shoreline change rates (Figures 28), the alongshore distribution of those rates, and a 10-m shoreface depth (Haller-meier 1977). Therefore, in recent times 60 percent of the sediments lost from the beaches appear to have moved in an alongshore direction primarily inshore of the ends of the Rudee Inlet jetties (Figure 24). Loss rates in the nodal zone area are based on 130 years' record and variations from one survey to the next were small (Figure 29), indicating that conditions have not varied as much there as elsewhere in the study area. Some of the unaccounted-for 40 percent of lost sediment may have been moved west by overwash or wind transport, or east and offshore into water that is deeper than the jetty ends. In addition, the static effect of sea level rise relative to land at 0.4 mm/yr (Hicks 1981) on a beach sloping at 1:30 would be a yearly loss of 26,000 cu m, or about 20 percent of the unaccounted-for sediment. Rising sea level may have had an additional, unquantifiable effect on the dynamics of the system.

Sound shoreline change

106. Dune construction, either by natural or artificial means, is usually accomplished at the expense of sand in the littoral zone. To compensate for the lost sand, the shoreface and beach profile, and, consequently, the shoreline, will retreat. This was probably the case following construction of the continuous dune between South Nags Head and Cape Hatteras which was begun artificially, using sand fences, between 1936 and 1940. Dune profile data and rates at which the dune grew are unavailable; however, if a final 5-m-high-by-60-m-wide dune with about a 3-m-high overwash platform resulted and a shoreface depth (i.e., the depth from mean sea level (MSL) to base of shoreface) of 10 m is assumed, the removal of that volume of sand from the littoral zone would result in a shoreline retreat of 11 m. Dune-building may be a factor in the increased shore erosion between Oregon Inlet and Cape Hatteras between 1917 and 1949 (Figure 45).

107. Driven by storm surges (Figure 15), overwash probably occurred frequently in the study area before dune construction; however, it likely had only a minor effect on the ocean and sound shorelines (Figures 52 and 53). Shoreline position changes do not seem to be related to island width for either the ocean or sound shorelines, except near existing or recently closed inlets. Away from inlets, the sound shoreline where the island was less than 900 m wide retreated at an average rate of 0.6 m/year (Figure 53), which is greater than the average retreat rate for island sections where the width was greater. Accordingly, overwash probably did not significantly affect the sound shoreline during the period from 1850 to 1980. If the effect were important, the sound shoreline at narrow places on the island would have likely prograded as sand moved from the beaches into the sound.

108. Away from inlets, the retreat of the sound shoreline can be accounted for mostly by sea level rise. At an average surface gradient of 1:100 near the sound shoreline, and a sea level rise of 0.004 m/year (Hicks 1981), the sound shoreline retreat rate would be 0.4 m/year, or nearly the actual rate measured. This rate will vary in the future as the sea level change rate relative to the island varies.

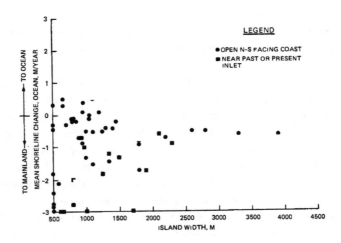

Figure 52. Ocean shoreline changes from about 1850 to 1980, Cape Henry to Cape Hatteras, as a function of island width in 1980 (shoreline changes are shown in Figure 28)

98

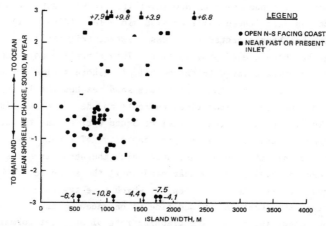

Figure 53. Sound shoreline changes from about 1850 to 1980, Cape Henry to Cape Hatteras, as a function of island width in 1980 (shoreline changes are shown in Figure 34)

Inlets and shore erosion

109. Inlets affect both sides of a barrier island or spit and have had a major impact on shoreline behavior in the study area. Shoreline changes that have occurred as a result of open inlets during the 130-year period of this study provide a basis to extrapolate shoreline changes caused by inlet processes backward in time to 1585 (Figure 9) and earlier. In some cases the effect of an inlet on adjacent shorelines is only one of a number of causes of the change in those shores.

110. Present inlets. Rudee Inlet, one of the two inlets presently open in the study area, is a small and stabilized feature that has only a small effect on adjacent shorelines; the recent history of Rudee Inlet is listed in Table 2. Oregon Inlet, unstabilized and many times larger than Rudee, is the only inlet that has been open continuously for the length of the study period. Since it opened just 4 years before the first shoreline survey was made, the survey data presented in this paper provide an excellent sequence with which to detail the inlet's behavior.

111. Oregon Inlet today is flanked by erosional ocean shorelines for about 8 km on either side of the inlet throat (Figures 28 and 31). Shore erosion, which is greatest near the inlet, decreases as distance from the

99

inlet increases. (The past site of New Inlet (Figure 9), just north of
Rodanthe, also has experienced major erosion since 1850 (Figure 31).) The
sound shoreline has been affected to a lesser extent (Figures 34 and 36), but
the net change has been one of progradation. This shoreline adjustment adjacent to Oregon Inlet is related to the normal alongshore sediment transport
(see paragraph 103) of beach sand. When this sand reaches the inlet throat,
some is carried landward by flood-tidal currents and deposited within the inlet system. The large shoal area in Pamlico Sound west of the throat at Oregon
Inlet is evidence of that inlet's trapping capacity. The sand composing those
shoals is coarser than the sound sands upon which the shoal area rests. Inlets such as Oregon Inlet probably trap sand until the sound shoals have grown
to attain a quasi-equilibrium condition, at which time the volume of beach
sand which enters the inlet on a flood tide is balanced by the volume carried
out on the subsequent ebb tide. The trapping rate of an inlet normally decreases with time after the inlet opens. However, when an inlet moves parallel to shore as Oregon Inlet has done (29 m/year on the average, Figure 38)
the entrapment rate may not decrease very rapidly because the flood-tidal
shoals never attain a quasi-equilibrium state of development.

112. An analysis of Oregon Inlet sand gains relative to adjacent ocean
shore sand losses provides an approximate means to illustrate that most of
the adjacent shoreline retreat is inlet-caused. Approximately 32,000 sq m/
year (4.2×10^6 sq m, total) of barrier island surface area has been lost
since 1849 within 8 km of Oregon Inlet (Figure 40) (to some extent, these
values have also been influenced by previously open New Inlet (Figure 9)).
To calculate the volume of sand moved, the depth to which the shoreface
profile has been modified must be considered; a reasonable depth (Hallermeier
1977) is about 10 m. Using the surface area lost (Figure 40) and the assumed
10-m depth to which erosion occurred, approximately 4×10^7 cu m of sediment
was lost from the barrier islands adjacent to Oregon Inlet between 1852 and
1980. The ebb- and flood-tide shoals in Oregon Inlet cover an estimated
2.5×10^7 sq m of Pamlico Sound. At an average estimated thickness of 2 m,
these inlet deposits contain 5×10^7 cu m of sands transported from the adjacent islands. Thus, according to this very crude analysis, the sands lost
from the beaches near Oregon Inlet can be accounted for within the inlet system, primarily in Pamlico Sound flood-tide deposits. Of course, superimposed
on the inlet-caused ocean shoreline change, is the long-term 0.8-m/year

retreat which exists for the entire study reach.

113. It is interesting to note that the sand entrapment rate has remained relatively constant since 1849 (Figure 40). The only perturbation occurred during the 1949-1963 period when the March storm of 1962 greatly changed the inlet (Figures 38, 39, and 40). Poststorm recovery, however, returned the system to its prestorm condition; Oregon Inlet is apparently still trapping sand (1980) at about the rate it trapped it in the first 66 years after it opened. As long as Oregon Inlet remains open and unstructured and continues to migrate south, the sand entrapment rate should remain near its 1852-1980 average value of 3×10^5 cu m/year. Adjacent ocean shoreline behavior should remain similar to that shown in Figure 28. As the inlet migrates south, the inlet-influenced ocean shoreline 8 km north and south of the throat also will migrate south.

114. Small, structured Rudee Inlet is presently not acting as a sand trap; littoral sand that is moved into the inlet throat is returned to Virginia Beach by hydraulic means. In the future this inlet will not likely affect adjacent beaches as long as present (1980) conditions prevail.

115. Past inlets. Inlets have been located, in historic times, in two regions: in northern Currituck Sound and centered around Oregon Inlet (Figure 9). Probably the largest prehistoric inlet (pre-1585), as evidenced primarily by beach ridges, was located at Kitty Hawk, North Carolina (Figure 11). Small ephemeral inlets have been opened during storms, but natural movements of sand along the coast have caused them to close within a few years. Only relatively stable passages through the barrier spits and islands are included in Figure 9.

116. Sands are deposited in flood-tidal shoals within the sound, on adjacent sound shorelines, and in ebb-tidal shoals in the ocean after an inlet opens. The net sand loss from adjacent beaches is reflected in an increase in the rate of ocean shoreline recession. Conversely, the sound at the inlet gains sand. If the inlet subsequently closes, the flood-tidal shoals frequently form a new shoreline or islands in the sound (Figure 37). Inlet closure is usually accompanied by ocean shoreline readjustment such that island width at the site of the former inlet increases; i.e., the ocean shoreline builds seaward.

117. An anomalously wide portion of a barrier island is often a clue to the previous existence of an inlet. In Figure 11, which plots island width

with inlet location and the length of·time the inlet was open, the anomalous island widths shown near latitudes 36°15' and 36°00' most likely reflect pre-1585 inlets. The existence of these sites indicates that the islands have existed in or near their present locations for at least the past 400 years.

118. Wide portions of barrier islands are usually less susceptable to a new inlet opening than are narrow portions. Thus, while the existence of an anomalously wide island reach often reflects the past site of an inlet, it probably is not a prime site where a new inlet will open. However, ocean and sound hydraulic characteristics, which were once maximized at the previous inlet location, probably did not change much; therefore, that general region remains a potential site for a new inlet. These sites can be identified in Figure 11.

119. Inlet effects on the ocean coast are rapidly muted after the inlet closes. Within a decade after closure, the effect of an inlet on the adjacent shorelines is no longer noticeable (see New Inlet, for example, in Figure 37). This occurs because alongshore sediment transport and the landward transport of ebb-tidal shoal material act to straighten the ocean side of the previously inward-flaired coast.

120. Conversely, the effects of an inlet on the sound shoreline may persist for hundreds of years (see Kitty Hawk Inlet, for example, in Figure 11). In the years after the inlet closes, the flood-tidal shoals may become islands, or may weld to the adjacent sound shores and spread and become less pronounced with time.

Capes and shoreline change

121. Cape influence is reflected in the behavior of adjacent ocean beaches. It appears that changes in the east-facing ocean shoreline at least 14 km south of Cape Henry and 10 km north of Cape Hatteras are dominated by the respective capes (Figures 28 and 31).

122. At Cape Henry the east-facing shoreline prograded while the nearby north-facing shoreline retreated (Figure 44), a situation that will likely continue into the future. The progradation could increase if additional artificial beach fill is placed on Virginia Beach. Some of the recently placed fill material moved north and was deposited along the east-facing shoreline (Figure 28).

123. The position of Cape Point at Cape Hatteras is highly variable (Figure 41), and its year-to-year movements do not appear to be predictable.

In general, though, the longer trend term appears to be to the south and west, as reflected in changes on the nearby shoreline (Figure 28). North of the cape, the shoreline movement has been one of retreat to the west, with the greatest westward retreat nearest the cape. West of the cape, the shoreline has prograded; this movement has occurred for a long time and is referenced in a large number of east-west-trending ridges. Future shoreline changes north and west of Cape Hatteras will likely be similar to those that have occurred in the past.

Shoreface-connected
ridges and shoreline change

124. Shoreface-connected ridges also appear to significantly influence the ocean shoreline in the study area. These linear ridges with a maximum of 10-m relief extend up to 10 km offshore from the shoreface in a northeast direction; side slopes are usually not more than a few degrees. Fields of such ridges are common from Long Island to Florida (Swift et al. 1972). Locations of the four shoreface-connected ridges along the east-facing ocean in the study area are shown in Figure 54 and listed in the tabulation below.

Name	Latitude
False Cape Shoal	36°33'
Oregon Shoal	35°52'
Wimble Shoal	35°33'
Kinekeet Shoal	35°23'

125. The ridges intersect the shoreface about 5 km south of some of the most prominant concave seaward shorelines in the study area (Figure 55). Except at inlets, these are the major sites along the east-facing ocean reach where the shore orientation varies greatly. The shoreline at and south of the ridge intersection is generally convex in a seaward direction. In all cases the site of the intersection is along a reach where the shoreline is rapidly changing from a northwesterly to a northerly direction.

126. Shoreline changes associated with the shoreface-connected ridges are predictable. Shorelines north of ridge intersections retreated, while those to the south usually prograded. One exception is south of Oregon Shoal where the shoreline retreated, probably because of the influence of Oregon Inlet. Shoreline changes adjacent to the ridge intersections appear to vary with time in a relatively consistent manner. Data shown in Figures 30 and 31 suggest the ridge influence is moving south.

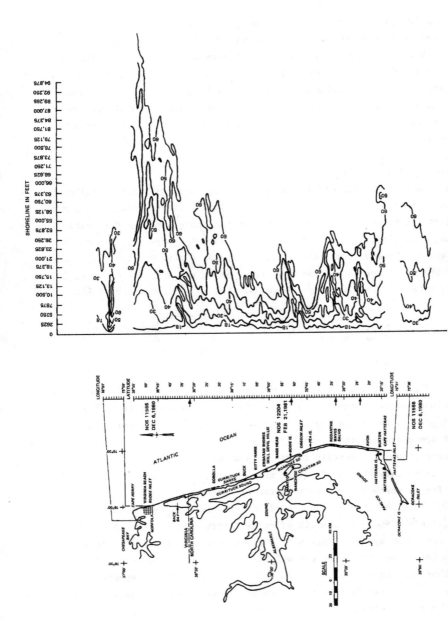

Figure 54. Bathymetry seaward of the ocean shore between Cape Henry and Cape Hatteras (the shore-normal scale is greatly exaggerated to show the alongshore variability in bathymetry; arrows identify sites where northeast-trending ridges intersect the shoreface)

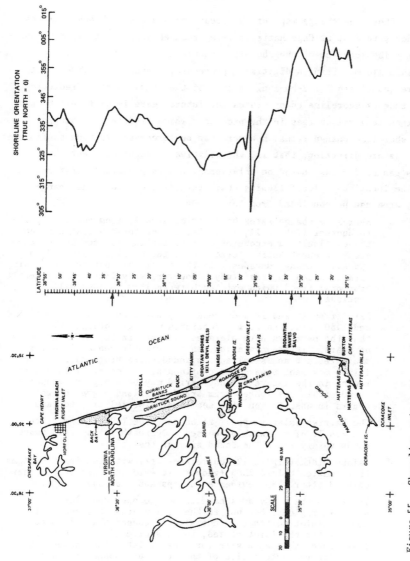

Figure 55. Shoreline orientation referenced to true north between Cape Henry and Cape Hatteras (arrows identify sites where northeast-trending ridges intersect the shoreface)

105

PART VI: SUMMARY AND CONCLUSIONS

127. Shoreline change maps of the ocean and sound shorelines from west of Cape Henry to west of Cape Hatteras were produced using historic NOS shoreline maps. The accuracy of the shoreline change maps is estimated to be at least within ±10 m. Using a digitizing procedure, average shoreline change rates were quantified for 2-km-long reaches of the study coast. Predicting the magnitude of shoreline change rates for future years is difficult because of undefined temporal changes in the processes which produce the changes. Relative shoreline change rates, however, can be forecast with some confidence in an alongshore direction; that is, the relative rates at adjacent shore locations can be forecast based on relationships with geomorphic features at or near the locations. The following characteristics of shoreline change in the study area can be concluded from this study:

a. Shoreline change rates have varied greatly from one time period to another (Tables 11, 12). Because of these variations and the difficulties encountered in attempting to account for them, accurate quantitative forecasts of the absolute magnitude of shoreline change decades into the future are not possible using data acquired in this study. However, very likely the general erosional trend which existed between 1850 to 1980 will continue.

b. Barrier spits and islands generally narrowed between about 1850 and 1980. This narrowing contrasts with geological evidence that the barriers have migrated landward in the past thousands of years. Island migration, in the classic sense, is ocean shore retreat and simultaneous sound shore progradation; i.e., island movement toward the continental landmass. Island narrowing in the 130-year study period may be a higher frequency trend within the longer term trend of island migration which occurs in association with sea level rise relative to land.

c. The barrier islands appear to be too wide (1980) to migrate as the result of overwash processes. Overwash-transported beach sands rarely reach the sound side of the islands.

d. Island width (Figure 12) correlates well with two inlet systems that existed before 1585 (geomorphic evidence) and inlets that existed after 1585 (evidence in maps and charts).

e. Inlets in the study area have tended to open and close in specific regions, but not in the same places in these regions. Because inlets often (but not always) caused the island to widen after the inlet closed, the historic inlet area became less susceptible as a site for a new inlet. But because the hydraulic characteristics of the ocean and sound caused the region to remain susceptible as a new inlet site, new inlets

106

tended to open near the sites of past inlets. Sites where inlets existed in the past 400 years (see Figure 9) are (1) Nags Head to Rodanthe and (2) Duck, North Carolina, to the Virginia State line.

f. Oregon Inlet, the only unstructured inlet that has been open in the study area for the entire study period, apparently affected the ocean coastline at least 8 km north and probably 8 km south of its 1980 location (Figure 37). Shoreline changes to the south were masked by the opening and closing of New Inlet. Shore erosion decreased exponentially away from the inlet (Figure 28). A rough calculation of ocean shoreline losses and Pamlico Sound sand gains indicates that nearly all the Atlantic Ocean sand lost from 8 km north and south of the inlet was deposited in Pamlico Sound. This net movement of sand in a westerly direction could, on a time scale of hundreds or thousands of years, be a major factor in island migration. Today, inlet processes are the major mechanism for moving littoral sand in a westward direction. Wind is probably second in importance.

g. Because of near-continuous southward migration of Oregon Inlet (about 29 m/year), the amount of littoral sand trapped in the flood-tidal shoals of Pamlico Sound appears to have been constant through time (about 3×10^5 cu m/year (Figure 40)).

h. Evidence of inlets that closed before 1585, most notably at Kitty Hawk, suggests the islands have not moved appreciably (i.e., not more than one-fourth the island width) in at least the past 400 years (Figure 12).

i. Capes affect adjacent beaches. In the past 130 years, the east-facing beach south of Cape Henry accreted, while the east-facing beach north of Cape Hatteras eroded (Figure 28). Concurrently, the north-facing beach west of Cape Henry eroded and the south-facing beach west of Cape Hatteras accreted. The net change is a very slight clockwise rotation and southward movement of the cape boundaries. The eastward progradation of Cape Henry and the southward progradation of Cape Hatteras are similar to longer term geologic changes in these areas as reflected in the orientation of beach ridges (Figures 2 and 7). Erosion north of Cape Henry and north of Cape Hatteras does not reflect past geologic changes; however, because these changes have occurred for at least 130 years, they can be expected to continue into the future.

j. A divergent alongshore transport nodal zone, identified using shoreline change data, exists near latitude 36°41'. This appears to be the only site of net alongshore sediment transport reversal along the east-facing ocean coast between Capes Henry and Hatteras.

k. Overwash probably has not been a major factor in producing changes in the sound shoreline. Most of the retreat in the sounds away from inlet influences can be accounted for by considering sea level rise on a gently sloping shore.

107

l. Shoreface-connected ridges intersect the ocean coast at four
    places. In each location, the shoreline north of the ridge
    intersection retreated, while the shoreline prograded south of
    the intersection. The ridge intersections are about 5 km south
    of the most prominent concave shore reaches (Figure 55) away
    from inlets. At and south of the ridge intersections, the
    shoreline changes rapidly from a northwesterly to a northerly
    orientation.

m. Characteristics of the shoreface-connected ridges are not
    dependent upon the net alongshore sediment transport direction.
    The False Cape Shoal is near a transport reversal; other ridges
    are located in areas where the net transport is to the south.

REFERENCES

Bagnold, R. A. 1941. The Physics of Blown Sand and Desert Dunes, William Morrow and Co., New York, N. Y., 265 pp.

Cumming, W. P. 1966. "North Carolina in Maps" (Text and set of fifteen maps), State Department of Archives and History, Raleigh.

Dolan, R. 1972a. "Beach Erosion and Beach Nourishment, Cape Hatteras, North Carolina," Dune Stabilization Study, Natural Resource Report No. 4, National Park Service, 20 pp.

_____. 1972b. "Maris Impact on the Outer Banks of North Carolina," Dune Stabilization Study, Natural Resource Report No. 3, National Park Service, 20 pp.

Dolan, R., Hayden, B., Rea, C., and Heywood, J. 1979. "Shoreline Erosion Rates Along the Middle Atlantic Coast of the United States," Geology, Vol 7, pp 602-606.

Dunbar, G. S. 1958. Historical Geography of the North Carolina Outer Banks, Louisiana State University Press, Baton Rouge.

Ebersole, B. A. 1982. "Wave Information Study for U. S. Coastlines; Report 7: Atlantic Coast Water Level Climate," Technical Report HL-80-11, U. S. Army Engineer Waterways Experiment Station, Vicksburg, Miss.

Everts, C. H. 1978. "Geometry of Profiles Across Inner Continental Shelves of the Atlantic and Gulf Coasts of the United States," Technical Paper No. 78-4, U. S. Army Corps of Engineers, Coastal Engineering Research Center, Fort Belvoir, Va., 29 p.

_____. 1981. "Changes in Changes Along the Coast," Paper given to American Shore and Beach Preservation Association Annual Meeting, Wildwood, N. J.

Field, M. E. 1973. "Report on Analysis of Offshore Seismic Records and Core Logs from the CERC Field Research Facility, Duck, North Carolina," Unpublished report, U. S. Army Corps of Engineers, Coastal Engineering Research Center, Fort Belvoir, Va.

Fisher, J. J. 1962. "Geomorphic Expression of Former Inlets Along the Outer Banks of North Carolina," M.S. Thesis, University of North Carolina

Goldsmith, V., Sturm, S. C., and Thomas, G. R. 1977. "Beach Erosion and Accretion at Virginia Beach, Virginia, and Vicinity," Miscellaneous Report 77-12, U. S. Army Corps of Engineers, Coastal Engineering Research Center, Fort Belvoir, Va.

Hallermeier, R. J. 1977. "Calculating a Yearly Limit Depth to the Active Beach Profile," Technical Paper No. 77-9, U. S. Army Corps of Engineers, Coastal Engineering Research Center, Fort Belvoir, Va., 22 pp.

Hayden, B. 1975. "Storm Wave Climates at Cape Hatteras, North Carolina: Recent Secular Variations," Science, Vol 190, pp 981-983.

Hayden, B., et al. 1979. "Spatial and Temporal Analyses of Shoreline Variations," Coastal Engineering, Vol 2, pp 351-361.

Hennigar, H. F., Jr. 1979. "Historical Evolution of Coastal Sand Dunes on Currituck Spit, Virginia/North Carolina," M.S. Thesis, The College of Williams and Mary, Williamsburg, Va., 121 pp.

Hicks, S. D. 1981 (Apr). "Long Period Sea Level Variations for the United States Through 1978," Shore and Beach, pp 26-29.

Ho, F. P., and Tracey, R. J. 1975. "Storm Tide Frequency Analysis for the Coast of North Carolina, North of Cape Lookout," NWS HYDRO-27, National Oceanic and Atmospheric Administration, National Weather Service, Rockville, Md.

Jensen, R. E. 1983. "Wave Information Study for U. S. Coastline; Report 9: Atlantic Coast Hindcast Phase III Nearshore Wave Information," Draft Technical Report, U. S. Army Engineer Waterways Experiment Station, Vicksburg, Miss.

Langfelder, L. J., Stafford, D. B., and Amein, M. 1970. "Coastal Erosion in North Carolina," Journal, Waterways and Harbors Division, ASCE, Vol 96, No. WW2, Paper 7306, pp 531-545.

Leatherman, S. P., and Fisher, J. S. 1976. "Quantification of Overwash Processes," Ph. D dissertation, University of Virginia, Charlottesville, 245 pp.

National Ocean Survey. 1981. "Tide Tables, U. S. Atlantic Coast," Riverdale, Md.

North Carolina Fisheries Commission Board. 1923. "Additional Inlets on the North Carolina Coast," Report on Special Committee on Inlets, North Carolina Fisheries Commission Board, _____ .

Shalowitz, A. L. 1964. "Shoreline and Sea Boundaries," V-2, Publication 10-1, U. S. Department of Commerce, Coast and Geodetic Survey, U. S. Government Printing Office, Washington, D. C.

Shideler, G. L. 1973. "Textural Trend Analysis of Coastal Barrier Sediments Along the Middle Atlantic Bight, North Carolina," Sedimentary Geology, Vol 9, pp 195-220.

Simpson, R. H., and Riehl, H. 1981. "The Hurricane and Its Impact," Louisiana State University Press, Baton Rouge, 448 pp.

Swift, D. J. P., Kofoed, J. W., Sandsbury, F. P., and Scars, P. 1972. "Holocene Evolution of the Shelf Surface, Central and Southern Shelf of North America," In D. J. P. Swift, D. B. Duane, and O. H. Pilkey, eds., Shelf Sediment Transport: Process and Pattern, Dowden, Hutchinson and Ross, Stroudsburg, Pa., pp 499-574.

Swift, D. J. P., Sears, P. C., Bohlke, B., and Hunt, R. 1978a. "Evolution of a Shoal Retreat Massif, North Carolina Shelf: Inferences from Areal Geology," Marine Geology, Vol 27, pp 19-42.

Swift, D. J. P., et al. 1978b. "Shoreface-Connected Sand Ridges on American and European Shelves; A Comparison," Estuarine and Coastal Science, Vol 7, pp 257-273.

Thompson, E. F. 1977. "Wave Climate at Selected Locations Along U. S. Coasts," Technical Report 77-1, U. S. Army Corps of Engineers, Coastal Engineering Research Center, Fort Belvoir, Va.

U. S. Army Engineer District, Norfolk. 1982. Phase I, General Design Memorandum, Rudee Inlet, Virginia Beach, Virginia," Draft Report.

U. S. Army Engineer District, Wilmington. 1980 (Sep). "Manteo (Shallowbag) Bay, North Carolina, General Design Memorandum, Phase II, Project Design," Wilmington, N. C.

U. S. Coast and Geodetic Survey. 1928. The Topographic Manual, Special Publication No. 144, National Ocean Service, Rockville, Md.

_____. 1944. "Photogrammetric Instruction No. 49," photocopy available from National Ocean Service, Rockville, Md.

Wainwright, D. B. 1898. "Plane Table Manual," In Annual Report of the U. S. Coast and Geodetic Survey, Appendix 8, U. S. Government Printing Office, Washington, D. C.

CPSIA information can be obtained
at www.ICGtesting.com
Printed in the USA
BVHW092314201118
533618BV00021B/2267/P